The Floor
of the Sea

Maurice Ewing and *Vema*.

The Floor
of the Sea

Maurice Ewing and the Search
to Understand the Earth

WILLIAM WERTENBAKER

Little, Brown and Company — Boston — Toronto

FIRST EDITION

T 11/74

Portions of this book appeared originally in
The New Yorker, in somewhat different form.

LIBRARY OF CONGRESS CATALOGING IN PUBLICATION DATA

Wertenbaker, William.
  The floor of the sea.

  Bibliography: p.
    1. Ewing, William Maurice, 1906–   2. Marine
geophysics. I. Title.
QE22.E9W37   551.4′608′0924   74-13064
ISBN 0-316-93121-7

Designed by Milly Robinson

*Published simultaneously in Canada
by Little, Brown & Company (Canada) Limited*

PRINTED IN THE UNITED STATES OF AMERICA

For Judy and Caleb

Acknowledgments

THE AUTHOR is grateful for discussions with many people who contributed generously of time, thought and experience, not only those quoted, whose help extended far beyond their quoted remarks, but numerous others whose contribution was no less great. All were helpful, many illuminating. Harriet Ewing, especially, gave advice and guidance. Discussions with Thomas Aitken, John Ewing, Marshall Kay, Captain Henry Kohler, Henry Moe, and Manik Talwani were particularly fruitful. Anthony Amos, Walter Alvarez, John Bastin, Allan Be, Enrico Bonatti, Lloyd Burckle, Ellen Coxe, John Dewey, Stephen Eittrem, R. D. Gerard, Arnold Gordon, Dennis Hayes, James Hays, John Heacock, Alice Hoffer, Lynn Isaacs, Adrian Lane, Marcus Langseth, Roger Larson, John Nafe, H. D. Needham, Dragoslav Ninkovich, Philip Rabinowitz, Vincent Renard, W. B. F. Ryan, Tsunemasa Saito, E. M. Thorndike, G. B. Tirey, Agatha Weston, Goesta Wollin and P. C. Wuenschel helped with the intricacies of Lamont or geology. Special thanks to those who gave original photographs or diagrams the author wanted. B. C. Heezen and T. E. Chase generously provided the endpapers. Lawrence Sullivan unearthed dramatic bottom photographs; and Marek Truchan found, and laboriously prepared, the dramatic, unpublished, profiler records.

The manuscript was critically reviewed by Walter Pitman and Maurice Ewing. The author is grateful for the many useful suggestions and corrections which they made; if any inaccuracy remains it is from his imperfect understanding and not their oversight.

Contents

Contents

Foreword

MAURICE EWING died, of a stroke, on May 5, 1974, a few days before his sixty-eighth birthday. He was a prodigious figure in a time when there were few such in any field. Two-thirds of the earth's crust has been explored in less than a quarter of a century, and to an extraordinary degree it was the work of Ewing and his students. He changed a subject from a polite academic backwater into one of the most exciting fields of enquiry being pursued today. How the earth's surface is formed, and deformed, has finally been understood. Ewing was a dedicated, but not anonymous, researcher, and tales of him were as numerous and widespread as his discoveries. During a decade or two men came from all over the world to learn his methods, which are now as much to be seen in Russia as in his own country.

Some visitors were taken aback that a chief method was dull, dirty work, and at the sight of the white-haired director of a leading laboratory clambering over the side of a ship to untangle greasy, rusty cables.

Though colleagues were awed by his researches, Ewing looked on himself as first a teacher. He was one who taught by example and partnership rather than by lecture or manifesto. As students he regularly took not the fleetest academic horses, as he well could have, but young people whom the academic system would ordinarily have rejected as poorly qualified or underachievers, and involved them in first rank discoveries at the forefront of the field — where they have remained. When, at sixty-five, he had to retire from the research group he founded twenty years earlier, Ewing started a new school and research establishment; he was planning its work when he died, while also looking forward to several months of research at sea for himself.

The Floor
of the Sea

One

The Floor of the Sea

MOST OF THE WORLD was first explored during the last twenty years. After the Second World War, a few curious men began to wander the oceans in small ships, profiling, photographing, and sampling the ocean floors. Little such work had been done before. The face of the earth was then almost entirely unknown beyond the edges of the continents. Beneath the sea — where nearly three-quarters of the earth lies — there is a landscape that no one had imagined. More than half the world is over two miles deep. So recently unseen and unknown (the first detailed sea-floor maps are only fifteen years old), this landscape has become very familiar to exploring marine geologists and geophysicists. For nearly a generation, ships and men prowled the earth, drawing the shape of a new world and finding new mysteries. Taken all together, these voyages of exploration make one of the great ages of discovery. They still go on. The submarine landscape is awesome and astonishing. Geologists once had declared it to be of no interest at all, but here, in just the last couple of years, have been found the forces that shape the continents, and the causes of mountain ranges, eruptions, and earthquakes — all of which had baffled geologists for over a hundred years. To those who try to understand the earth, the present period is comparable — and is compared frequently by geologists nowadays — to the years after the discovery of radioactivity and relativity, or the origin of species. Indeed, geologists have reached as far back as Copernicus for an adequate comparison.

In the shadow of the abyss, clefts, cliffs, plains, and mountains lie in uneroded grandeur. Plains flatter than the eye can measure stretch over the curve of the earth. Ragged hills spring from the plain and cluster on higher ground. Seamounts rise out of the landscape, their flanks sheer and unencumbered by foothills and ridges. Gaping chasms drop as deep again as the sea floor itself. Massed peaks rise rank on rank into a chain of mountains greater than any above the sea. The continents rise like mesas from the ocean basins. Were the earth not so segregated, it would be entirely covered in ocean over a mile deep. This segregation, besides being agreeable to us, is fundamental — the ocean basins are not simply depressed continents but are made differently from them. It has been one of the most essential things known about the earth since Maurice Ewing and Frank Press, who was then Ewing's student, established it in 1949. It is what Ewing liked to call a brutal fact, and the sort of thing he liked to fling in the face of theorizing geologists. It took some years to fully digest and understand; and another brutal fact is that no one yet can really explain it, though it is a central assumption of the new understanding of the earth that has developed since the late sixties. Curiously, where geologists once wondered why there are ocean basins, now there are difficulties in explaining continents.

That the floors of the oceans are no longer terra incognita is largely because of Maurice Ewing's curiosity, which was relentless. "Ewing has, directly, by his influence, and by his challenge," the head of another research laboratory has said, "oh, something like tripled the rate of our understanding of the earth — above the ordinary explosion of knowledge that has come to every science since the war." What Ewing did, and what he caused to be done, are prodigious but, equally, part of a busy and rapidly growing field of inquiry. Twenty years ago, no man in oceanic research had to be introduced to any other, wherever in the world he was from. Today, men from the same state don't know each other. Hundreds of research papers are presented at annual scholarly meetings — the work of hundreds of scholars, hundreds of graduate students, dozens of countries, and dozens of ships, with hundreds of men to man them and tens of millions of dollars to support them, all scrutinizing the ocean floor. Yet just a few years ago it could be said that at least half of the geophysical information about that part of this planet was gathered by Ewing and his research group.

William Maurice Ewing — he went by his second name, and pronounced it in the English, rather than the French, manner — was a soft-spoken, tired-looking man, a Texan by birth, a physicist by training. He was sixty-seven when he died in May 1974 and had spent a large part of his life at sea, exploring the earth. He liked to spend a couple of months a year at sea, and was apt to fly off at odd moments to join a ship in Manila or Punta Arenas. Through his trade, he knew distant and out-of-the-way ports, and stretches of ocean that merchant seamen never travel. He was a big man, over six feet tall — a large, rough-looking character, he once said — and broad-shouldered from hauling cases of dynamite and other geophysical equipment and parts of ships. He had a dense mane of white hair, of which a shock sometimes fell over his eyes. His face was broad, lined, and weathered, and his eyes, under puffy lids, were little and blue. Ashore, he sometimes peered through a small pair of silver-rimmed spectacles. He looked kind of grumpy, and some people were scared of him. Ewing's life was a source both of constant pleasure and of constant frustration to him. "In this business," a colleague of his says, "you can be like a kid in a candy shop." Ewing plundered the shop for forty years with formidable ingenuity. For him, the world was full of goodies, if he could only get around to them. He spoke of the ocean floors with as much familiarity, regret, and frustration as some other passionate traveler might the antiquities of Italy. Ewing wanted to understand how the earth works. He wanted to very badly most of his life. In recent years, he had five hundred people and two ships to help him work on the problem. He didn't appreciate anything that got in the way of learning about the earth. With Ewing something was either a profitable occupation or a total waste of time, but he always had more to do in a day than can be done in several. He could not wait to learn anything, wait for someone to invent a better instrument, wait until tomorrow to learn if his ships at sea discovered something today, wait until Monday for Saturday's mail. It caused him great anguish to let an unfamiliar stretch of ocean go unexamined, or a few unscheduled hours of ship time go unused, or to sail at less than full speed while instruments were recording bottom. "There are other men who think in worldwide terms," a geologist has said, "but Ewing — sending his ships around the world over and over, like no one else — has also done in worldwide terms."

In 1949, Ewing founded Lamont (now Lamont-Doherty)

Geological Observatory, which has become the leading marine geophysical research center in the world. (In 1972, after more than twenty years as director of Lamont and Higgins Professor of Geology at Columbia University, of which Lamont is an offshoot, Ewing left to found a new research center in Galveston, Texas.) The observatory is on a country estate on the Hudson Palisades a few miles north of the George Washington Bridge. Among the trees and hedges and gardens are some attractive old buildings — in one handsome pink sandstone and clapboard house was Ewing's chart-strewn office, to which, as staff member and emeritus professor, he returned occasionally for stints of research — and half a dozen sprawling new laboratories. Lamont's research grants amount to more than ten million dollars a year now, and besides its staff at Palisades it has people at sea and in small outposts scattered around the world; but to describe the original size of Lamont, Ewing would solemnly raise one finger. Lamont is very much the lengthened shadow of Maurice Ewing. Lamont keeps its two ships, *Vema* and *Conrad*, each at sea three hundred days of the year, for which they have been called the hardest-worked research vessels afloat (someone has said they circle the earth as constantly as two moons); was responsible for sea-floor photography and geophysical operations on the National Science Foundation's Antarctic research ship *Eltanin* for eleven years until she was mothballed in 1972; and is a founding member of the Joint Oceanographic Institutions Deep Earth Sampling program — JOIDES — whose drill ship *Glomar Challenger* has been circling the earth since 1966. Every few weeks, a sheaf of data about some part of the world that no one has seen before comes back to Lamont. The observatory monitors the earth's murmurs and arhythmias with the largest seismograph station in the world, and it even put a few of its stethoscopes on the moon.

In personnel, ships, and income, Lamont is still the smallest, as well as the newest, of the three major ocean-research groups in this country (the Scripps Institute of Oceanography and Woods Hole Oceanographic Institution are the others; sometimes, including Miami University, people in the field speak of "the big three and a half"), but according to a report to the government, whence comes most research money, Lamont has "produced over half the data of a geophysical nature" about the ocean floors and "initiated all . . . except two" of the ways of getting that data. Lamont is a formidable power in the earth sciences. Its artillery is always alert to the publication of new theories, and subsequently may lumber around into position and lob a few pertinent facts from its capacious larder over on the theory's author, sometimes demolishing the theory, occasionally establishing it. Ewing, too, was a force men reckoned with in each of the fields in which he worked (because he was interested in the earth, not fields, he worked in many), and he is sufficiently a legend in the earth and ocean sciences that he is spoken of with respect, if not awe, for his inventiveness, his energy, his impatience, and his wit even by scientists who never met him. He was not particularly well known except to other scientists. He may have been particularly poorly known. Fame in science today is related to Nobel prizes, and Ewing never had one. There is no Nobel prize for the investigation of the earth — for geology or geophysics, or marine geology and marine geophysics (which were just a-borning when he began practicing them), or seismology, or any of the other overlapping fields and subfields he practiced in that are lumped together generally as earth sciences — though it does not seem unlikely that if there were he would have had it, for any of several pieces of work. Ewing did win the National Medal of Science, in 1973, and the Vetleson Award, which some regard as the equivalent for the earth sciences of a Nobel (though a few scientists felt Columbia University — the administrator of the award — was a little insensitive awarding it to its own boy before anyone else, richly though he deserved it).

Although Lamont has been exploring and mapping the earth for more than twenty years and Ewing for longer, confusion about what they do has refused to die out even among some of their close neighbors, despite the cauterizing effect of Ewing's tongue. There are some in the surrounding county who believe that the observatory looks at stars, and confuse it with the white domes of a nearby antimissile radar installation. Others suppose, more knowledgeably, that with its ships Lamont studies the sea; a magazine of wide international circulation has called Ewing "dean of American oceanographers." Brightly encouraged to say how fascinating the great ocean is, the "dean" snarled: "The ocean's a murky mist that keeps me from seeing the bottom." He was not an oceanographer and it would have been a great help if the whole thing would dry up. A few wise souls know that the bottom's the thing; they show their sophistication by dropping the name of a well-known skin diver. But the amount of ocean bottom that can be reached by diving — by skin or by sub — is trivial. The floors of the ocean basins average three miles deep; even modern, "space-age" (with-it oceanographers have taken to calling their realm "inner space") submersibles offer little at these depths. This, too, Ewing explained. "Friend," he said to a curious soul, after buying his first (surface) ship, "in my trade, diving's like sitting on top of a flagpole, and there's not very much you can learn about better at the top of a flagpole than on the ground." So Lamont studies the bottom from the top, the earth from the sea, with dredges, coring tubes, drills, thermometers, cameras, and sonar; with gravity meters that respond to changes in the earth's consistency; with magnetometers; and with seismic reflection and refraction, which are to sonar about what an X ray is to a camera.

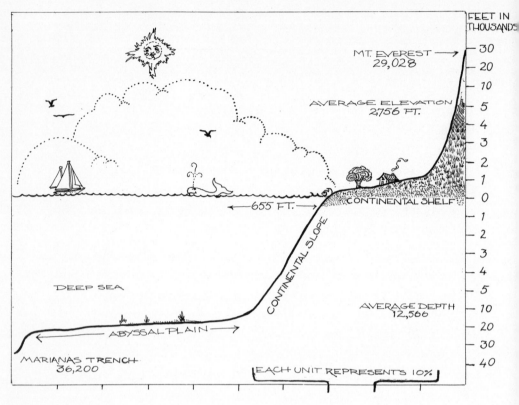

Amount of the earth's surface at different depths and elevations.

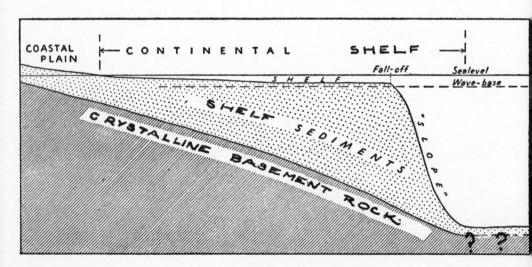

The continental shelf and its underpinnings.

Some people were deeply offended by Ewing — to the even deeper delight of others. Oceanic research was long a gentleman's science, probably because most scholars could neither maintain ships nor get others to, and Ewing did not follow all the rules while making his discoveries. Says an associate: "I was with him once at an oceanographic congress, in Brussels, and they were all there, all the distinguished oceanographers of the world, with their brandy and big cigars, telling each other of the hardships they had faced, and the setbacks and limitations they had had to accept at sea — particularly from the commercial depth sounders of the day. And there was Ewing, and he was trying to explain how if they would only take a screwdriver and a soldering iron they could fix the instruments. And they didn't like it one bit, I can tell you." Personal stakes aside, it is agreed by some oceanographers (while others maintain just the opposite) that inventive intellectual work is impossible in small ships at sea and that even carrying out a planned line of action becomes, after about two weeks, almost overwhelmingly onerous. Ewing, if even diffidently approached by such frailties, was still after several months at sea more thoughtful, energetic, and alert than some scientists ashore.

Ewing was curiously gentle when he was not desperate, though he was desperate to learn something most of the time. His expression was frequently worried, his shoulders stooped, and his gait shambling (he had some limp in his walk ever since he was washed off a ship during a midwinter gale in 1954). His mood usually depended on how much of the day he spent looking over data and how much on interruptions from the telephone, on the likelihood of his soon having two months at sea, and on his success in avoiding embroilment in administration. Paperwork was not rated a profitable occupation by Ewing, and he always was — his words — unsystematic and swamped. Wherever he went, an old friend fondly observes, you could see the bow wave of confusion out in front. There is no question that he was swamped, but unsystematic is harder to prove. Among the evidence to be weighed is a number code he devised for ships' masters and chief scientists to report by radio their daily results, giving him important findings in time to order changes of schedule while expending a minimum of money on the routine. He was not alone in accusing himself of bad administration; but whatever the administration was, even outside of Lamont people have remarked that building a major lab from nothing is not a bad argument for it. Yet Ewing's resignation was probably welcomed by a number of Columbia officials. (He already had been told that compulsory retirement would be honored in his case.) Ewing's demands were frequently called outrageous by those who were called on to supply them; he always wanted more than anyone ever did before, rousing cries of pain and outrage. In the end there usually was more for everyone. But he resigned, he said, because of a new university administration's desire to reinterpret the terms of endowments, a policy that had resulted in one donor's withholding payments amounting to several million dollars. "Policy," Ewing snorted, "is sabotage, originating in higher echelons."

Lamont and Ewing have had a reputation in Washington, according to one bureaucrat, of supplying "more research for a dollar than anyplace else," but Ewing always insisted: "Professor of geology is my real appointment here; the other junk I have to do. The chance to do science is my pay." With that he was a big spender. It was a rule of thumb at Lamont that if the light in his office was out before eleven, Doc — as his students addressed him and almost everyone else referred to him — was unusually tired. To avoid interruption, he would arrange to meet graduate students with a thesis in progress from ten at night until two (he occasionally dozed off from time to time), or at seven-thirty in the morning, times he would otherwise have given to his own research. To see much of Doc it was sometimes necessary to go to sea with him. But at sea for thirty-day stretches he was up every hour or so looking at the instruments to be sure the ship wasn't steaming past something interesting. One of the first things some young geophysicists learned about the earth was that they couldn't keep up indefinitely Doc's pace of doing science — and he could. This made his point of view a little special. "Once when I was a student," says Charles Drake, a prominent geophysicist, "Doc joined our ship at Nassau. He looked exhausted. He probably was — he usually is. I took the first watch and he the second. He looked so pooped that I let him sleep that night, and stood his watch, too. Along about 4 A.M. down comes this real mad body. 'Why didn't you wake me up?' 'Well, you looked tired.' 'Don't you ever do that again,' he yells, and bow-wow-wow-wow-wow-wow." Away from Lamont and his duties as director, Ewing worked equally hard on science. A few years ago at a conference in Buenos Aires on southern hemisphere matters, a scientist who probably should have known better was surprised to find him in his room during a reception, hard at work on a paper, with charts spread about the floor. He was nonplussed to find that the paper, far from being scheduled for delivery at the conference, concerned a place called Resolute Bay, eight hundred miles northwest of Hudson's Bay. Ewing was just collecting his pay.

From time to time, Ewing envied his friends' relaxation, even as they envied his application. Several years ago, inspired by a colleague who piled his family into a camper and drove them through the national parks on his way to a scholarly meeting in California, Ewing piled his wife and their four children into their car for an extended spin about the countryside.

There is no evidence that Lamont, or any of its ships at sea, functioned less well, or overlooked any major discovery in his absence, but after a day he was so intolerable to everyone, including himself, that he flew back to New York to recover his peace of mind. "The first day out," says his camping friend fondly, "he'd sell the trailer." Friends maintain that — with the possible exception of vacations — Ewing did nothing that he did not do well. He approached everything with the same intense concentration. When it was focused on the earth, he was said to walk through whole roomfuls of friends without seeing one of them. The same friend said: "When he's preoccupied he's irritable and hard to reach, and people think him surly; but when he turns himself on to social life he can be a most engaging man, and bend over backwards to help out or be pleasant and charming." A neighbor said: "Maurice is not only large, physically attractive and masculine, when he decides for a few minutes to turn that five-hundred-watt mind on you, it's difficult not to feel flattered." One of his oldest associates delights in telling of a dinner party where Ewing was seated beside a man who was studying, say, wombats: "He felt his research was at an impasse, and Maurice listened through the soup, the meat, and the salad while he described it, and finally said, 'Have you tried this?' 'My God,' the man says, 'that's it.'"

Ewing's intense preoccupation with his work probably affected his marriages. He married three times. As an undergraduate he met Avarilla Hildenbrand, and they were married five years later, in 1928, when he was in graduate school. They had one son, William (he was killed in 1963, when an Air Force plane in which he was riding crashed). After Ewing began doing war work in 1940, and was increasingly long away from home, his wife returned to her family in Texas and later divorced him. In 1944, he was married to Margaret Kidder, whose family summered in Woods Hole. A Bryn Mawr graduate and the author of some plays, she had been living in France when the war began. They had four children, Jerome, Hope, Peter and Margaret. The marriage ended in divorce in 1965. "It's not that Maurice didn't care," says a friend. "Twenty years is too long. But, especially during the war, he'd get immersed in some project or be off at sea, and his family would see little or nothing of him for long periods. Then when he was around he was very demanding. I don't think Midge ever got used to the amounts of food he expected — like chops for breakfast instead of melba toast." In 1965, Ewing was married to Harriet Bassett, a small, blond woman with eyes of a startling pale blue. At Lamont, she ran the director's office and kept in touch, and she continued this function in Texas. A visitor could get from her an excited and swift summary of the latest discoveries; there was little about the operation that she did not know.

Things as indefinite as taste and intuition in the end determine a scientist's quality. Many tireless men, and many of seemingly awesome analytic power, have spent themselves on matters of little consequence. Ewing's scientific intuition was a favorite subject of his associates. To one man, his quick opinion about any problem was worth more than anyone else's long considered judgment: "His insights are so astonishing, his fund of information tremendous. His judgment is infallible on anything except the amount of time needed to do everything — which he grossly underestimates." Not the least part of any research is deciding what will repay effort. "Ewing was always right there with the most important thing to do," says George Sutton, a former student and colleague who continued to work with him now and then on moon seismology. "He'd never touch a subject where you had to go three more decimal places to add anything." A colleague calls Ewing a genius (by way of

apology — "You can't be a genius without antagonizing people"), other people different things. "A cream-skimmer," a very
eminent oceanographer said, sniffily, some years ago. Redolent
of the disapproval of ancient nannies, the epithet would have
had its subject's enthusiastic concurrence — one of the annoying things about him. "He's moved in and out of a dozen fields,"
says John Lamar Worzel, a former associate director of Lamont,
with whom Doc worked on gravity, underwater cameras,
and marine acoustics, "and other people have been working on the details ever since." Says a friend: "He discovers
problems, and methods of solving them. Other people finish
them." One problem Ewing created fifteen years ago was de-
signing a seismograph to go on the moon. He gave it to Frank
Press, who gave it to Sutton, who gave it to Gary Latham, a
Lamonter who accompanied Doc to Texas and is NASA's principal investigator for lunar seismology; there are four seismographs on the moon. Where some spend a lifetime learning
about gravity, magnetic fields, underwater sound, rocks, seismology, the atmosphere, sedimentology, paleontology, ice ages,
the earth's inner structure, or the structure of the continents or
ocean basins, Ewing insisted on studying them all, boning up
nights on the background of new fields like any graduate student, because all, even the moon, hold clues to understanding
the earth. Nonetheless, Ewing's work in each of five different fields would have made him an eminent scientist, even had
he done no other, according to Press, who is now the chairman
of MIT's Department of Earth and Planetary Sciences.
Ewing's first professional paper, a combination of natural observation and classical physical theory, published in *Science*
when he was eighteen and a junior in college, was called
"Dewbows by Moonlight." Not every paper since sounded as
romantic, but the impulse was. "For my taste," Ewing said,
"the most exciting thing you can do as a physicist is investigate

the earth." He was, one scientist has said, the first to do geological work of any consequence at sea. Imagination is seeing what is not agreed upon. He had the knack of looking where there was said to be nothing and finding something exciting. So little was learned of the land beneath the sea before Ewing began to look about that generations of oceanographers and geologists had concluded that it was featureless — but it was featureless only in their minds. "There were very few geological scientists then," according to one of them, "who had noticed that the oceans made up most of the world, and they didn't take that part of it into account in their theories. There was a sort of continent syndrome. Ewing's was a perceptive commitment, and a great one." Today there is intense interest in the sea floor, with much research money and academic fame to be had for the study of it. And some explorers have set out, like the Spaniards of old, to discover what they have been told is there. Few men, Robert Frost remarked, are anxious to be the first anywhere, but many are eager to be the first to be second.

Two

The Chief Scientist
Dreamed of Cervantes

EWING GREW UP on the plains of north Texas, in Lockney, a community of small farms and small ranches on a spur of the railroad northeast of Lubbock; later the family moved a few miles to the smaller town of Tulia, up toward Amarillo. The Texas panhandle is a parched country of dusty homesteads and dusty towns with one street, several stores, a few houses, and a grain elevator. The Ewings were independent, but the land was not generous. An early attempt at homesteading, during which three infants died, failed for lack of water. At Lockney, they brought water to their farm by wagon — though drought did not predestine Maurice to the ocean. He was the oldest of seven children, of whom only one daughter, who married early, was not sent to college. An acquaintance recalls a friendly mob ruled by two strong characters. Hope Ewing was a small, energetic woman with high ideals, Floyd Ewing at once easygoing and exacting, taking charge without precedence. Maurice played trombone in the school band and got a less-than-middling education. "My first idea of what I was going to be came out of *Popular Mechanics* pictures and correspondence-school ads showing the man pulling the switch," he says. But when he set out to learn to pull the switch, the good colleges he applied to told him to go to cram school for a year and apply again — earlier. He had a ticket to the only place that would take him as he was, and had packed his trunk, when one of the good schools — Rice Institute (now University) in Houston — wrote again, this time accepting him. He went there, in 1922, aged sixteen. Rice gave him a medal of honor thirty years later — Rice does not believe in *honorary* degrees — and he heard the reason he had been admitted. His high school math teacher had written indignantly, but apparently convincingly, backing him over "anyone I ever met, or I guess anyone you ever met." Ewing said: "Under his guidance I had changed from a person who could do no math problems to one whose favorite thing was algebra." Though that was a good thing, it still was a while before he could understand his calculus teacher at Rice, a few steps, like trigonometry, being customary between algebra and calculus. As a freshman, he said, he worked from five to midnight in a drugstore, and had classes and laboratories from nine to four. But he won a scholarship, and for the next three years worked in the library and graded papers. Somewhere in the schedule he was first trombone in the marching band, a chair he held through graduate school. He was a big, lanky boy. Summers, he worked first in grain elevators, later prospecting for oil. There is probably more oil than water under north Texas, and such applications of physics as the torsion balance and explosion seismology were then the very last cry in oil prospecting.

Ewing was always excited by his education. When he talked

about Rice in his study at Lamont, surrounded by cruise reports, charts, and data books, his voice rose and quickened, and he pawed through books until he found three commemorative volumes of Rice's founding ceremonies in 1912, with a list of famous scholars attending and speeches on learning by eminent philosophers and scientists. Rice had been founded with the highest intellectual ideals, and he had felt them and was always impressed. "For two years I majored in electrical engineering," Ewing said. "And then I began to suspect. The engineering students I had noticed used funny paper with red lines down and across it. They were careful how they made figure fives. An inkblot was a catastrophe. The professors were sarcastic Yankees. But the three men I'd been learning from, in physics and mathematics, were such gentle people, the kind of people I liked. For half a year, I studied the upperclassmen. The engineers were all worried. There was a group of us, all sophomores, living in the same entry; most said they agreed physics was more interesting, but wondered how you could make a living at it. I decided to change anyway. I was already working my way through. Nowadays, physics isn't a bad career. And the nicest people I ever met, before or after, taught physics. The blessing of Rice was two math professors with Harvard degrees fresh from Sheldon traveling fellowships to Edinburgh, Paris and Rome, and the physics professor, who was from Cavendish lab at Cambridge. I had essentially as private tutors men straight from the greatest centers of learning in the world. There was an excitement that lasted clear up through the years I taught at Lehigh. Up through now. I think I am the most grateful person who ever went to Rice.

"At the end of my sophomore year, Professor Ford, one of the math teachers, asked me what I was going to read during the summer. Read? I hadn't thought. Working in a grain elevator is long hours. He said, 'It's as I expected, so I got something out for you.' Students couldn't take library books for the summer. He had three volumes of Faraday's *Experimental Researches*. I thought so highly of him that I didn't want to look appalled; and I read them, I think. Once, that autumn, the philosophy professor, making conversation, asked what I'd read recently. I told him, and he still tells me about it sometimes when I see him.

"The physics professor was kind of dictatorial. If he gave up on a student he'd say, 'Humph! you'd better take chemistry.' He ran very good colloquia in which current papers of the literature were presented. Graduate students and honors students were expected to attend. You'd have tea Cavendish-style and hear a paper and discuss it. There were many distinguished visitors we got to meet and have tea with who I would otherwise have thought not mortal. We were in the midst of the great revolution of quantum physics; all the classic papers appeared while I was at Rice. There were many eminent places where you didn't hear about these things at all in physics class; we had the papers as soon as the library had them. We discussed them, and the professor had us do all the experiments, like the diffraction of electrons. Not many of them worked. He himself had done a rough version of Millikan's Nobel prize experiment — measuring the charge of the electron — years before Millikan, and he considered himself a sort of black sheep, the only one of the Cavendish group, including some of his brothers and in-laws, who didn't have a Nobel prize. Well, you don't get measles except from people who have them, and I sure was in contact with people who had bad cases."

In 1930, Ewing, by then incurably infected, went to teach at Lehigh University in Bethlehem, Pennsylvania. He was crazy to do research, he says. But there were certain hazards. Science had not become a public utility. Scientists financed their research with money that today — one has said — wouldn't tempt them to Washington for an afternoon meeting. Ewing taught eighteen class-hours — later raised to twenty-one — of elementary physics, for which, of course, there were the additional duties of preparation, grading, and lab work. No doubt in consideration of the Ph.D. and wife he had acquired during graduate school, he was paid eighteen hundred dollars a year, in nine installments. Inevitably, he taught summer school. He was quite free to research on his own time, and with his own money and equipment. Out of the geophysical resources of the Lehigh Valley he extracted whatever research he could, a mixed bag of monographs — "Locating a Buried Power Shovel by Magnetic Measurements," "Seismic Study of Lehigh Valley Limestone," "Magnetic Survey of the Lehigh Valley" — with implications for basic problems in pure science. When quarries were blasting, he and his students thriftily recorded the seismic waves along the valley. One flush winter when they were given explosives of their own, they studied the propagation of explosive waves in a solid medium horizontally bounded by a liquid and a gas — the ice sheet of a frozen pond. But when a strip-mining company offered a fee larger than Ewing's salary if he would trace a coal seam geophysically, he refused, until it was agreed that he could publish his results — and sacrifice the fee. Every year he had a paper to read before the meeting of the American Geophysical Union.

One afternoon in 1934, when Ewing was still only an instructor in physics, he had callers in his basement office at Lehigh. "It was a significant and fruitful meeting," he recalled. "I remember it like yesterday. It was a slightly snowy day in November, and two well-dressed gentlemen entered whom I barely knew — Professor Richard Field, of Princeton, and Doctor William Bowie, of the Coast and Geodetic Survey.

They were both sort of imaginative and enterprising people. They had known about me because I came to meetings of the American Geophysical Union every spring and read a paper. They wore derby hats, and coats with fur collars — I remember the snow on them. They said they wanted to interest me in the study of the continental shelf. They thought it was a very important geological problem to see if the steep place where the shelf ends was a geologic fault or the result of outbuilding of sediment from the land — was it a basic geologic feature or a superficial appearance? And they wondered if seismic-refraction measurements, such as I had been working with, could be used. I said yes, it could be done, if one had the equipment and ships. If they had asked me to put seismographs on the moon instead of the bottom of the ocean I'd have agreed, I was so desperate for a chance to do research. Bowie and Field thought the Geological Society of America would back me — 'especially,' they said, 'if you don't mention us.' I got the grant, for two thousand dollars, and my chief efforts from that day to this have been devoted to the problem those two great men propounded." Ewing could not look at a question without seeing a dozen more related ones, or be satisfied without knowing if what he had seen was typical or exceptional; once begun, he was led by stages to look at the entire world. Bowie and Field thought it would be nice for geologists to know where the edge of the continent — of any continent — is. Was it the edge of the continent shelf, or was that just the edge of a great pile of mud? So they asked Ewing to find out how deep the sediments were there, just offshore. The question is easy to state, but if geologists could see beneath the surface, a lot of their questions would have been answered years ago. Getting an answer to the question posed by Bowie and Field required some sophistication. Ewing's method, explosion seismology, had never been used at sea before he began his survey in 1935;

he, and others after him, have since made it one of the most revealing ways of studying the earth at sea.

Seismology — the observation of vibrational waves in the solid earth — began about the middle of the last century and has led to most of what we know about the earth that we cannot dig up. Waves from explosions were first used in 1848 by Robert Mallet, an Irish geologist and early student of earthquake waves. Earthquakes and explosions start a number of different kinds of waves in the earth. Physicists call them elastic waves (as in *Elastic Waves in Layered Media* by Ewing, Jardetzky and Press, a classic in the field). There are two classes of them, body waves and surface waves, the former being given to making beelines through the earth, the latter content to come around on top. Most seismology, including seismic exploration, used the body waves, of which one kind is simply sound waves. The waves in the earth are generally in octaves too low to hear, but witnesses have reported hearing the earth humming and groaning during great earthquakes. From the body waves it was possible, for instance, at the turn of the century, to get indications of something now taken for granted — that the earth has a core. The primary body waves, or P waves, will travel through anything, while the secondary body waves, or S waves — besides being slower than the P waves — can only penetrate solids. The inability of S waves to travel through, or even very near, the center of the earth, is evidence for a core, a liquid one. The size of the core was found, in 1914, from the shadow it casts in S waves — it's eighteen hundred miles down. (For a while the core-mantle boundary was called, noncommittally, the Oldham-Gutenberg discontinuity.) Similarly, the "discontinuity" in seismic waves about thirty miles below the surface, discovered by the Yugoslav Andrija Mohorovičić, and named for him, was the first measure of the boundary of the crust we walk on and the mantle layer between crust and core. In 1848, Robert Mallet was measuring the speed of sound in rock. He exploded gunpowder in granite and, through a telescope, watched the tremors in bowls of mercury he had set a distance away. Today, the job is done with TNT, radiotelephones, and seismographs. Oil companies took up explosion seismology in the nineteen twenties, and Ewing extended the technique to the oceans. Explosion waves are reflected by the strata immediately beneath the surface, and with enough measurements the geologist can construct a sort of subsurface topographic map, in which the petroleum geologist, for example, can trace oil-bearing strata to new drill sites. Waves also get bent, or refracted, along strata for a bit and can be made by complex calculations to reveal the speed of sound in each layer — the speed of sound in familiar substances like granite or limestone now being known.

Off the coast of Virginia, where Ewing made his survey, the continental shelf is some seventy miles wide, deepening very gradually to about six hundred feet at the edge. The effect of a layer of water on seismic measurements was uncertain. In swamps and bayous, oil prospectors pushed their geophones into the bottom with long poles, and the water over the continental shelf was still shallow enough for equipment to be lowered easily to the bottom on cables. "I begged some obsolete equipment from an oil-hunting outfit I'd worked for. We were to go on the government ship *Oceanographer*, which was engaged making topographic surveys out of Norfolk. That determined our location. The plan was to run a line of seismic stations across the land, starting near Richmond, where the basement — the bedrock — outcropped, and going across the coastal plain to the shore, tracing the basement below. In Fortress Monroe, there was a water well to basement, about two thousand feet deep — the commandant was alert to the dangers of conquest by siege — and the driller's log would give us an independent record of the strata to check our figures against. We achieved a very close agreement. On the ship, we would take up the line of stations from the shore to the edge of the shelf, seventy miles offshore. To fit her schedule, however, we had to do that work before we did the land work. The written orders to the ship's captain were to allow us on board and — this for the eyes of congressional busybodies — do experiments for us on a not-to-interfere-with-ship's-work basis. Explosives were used occasionally in the course of the ship's work, but it was understood that the orders were to be interpreted liberally." The ship sailed shortly after the last day of classes at Lehigh.

Important as the project was to him, Ewing could not get his obsolete equipment ready and meet all his classes. He had two colleagues, Albert Crary, then one of his first graduate students and now a divisional director of the National Science Foundation, and H. M. Rutherford, who was in charge of a seismograph station in Pittsburgh. The three men worked all night the last day of school, and drove off tired in the morning. Ewing had a panel truck that had been discarded by the local power company; the tires were bad, and it was overloaded. It looked the chosen vehicle of the archetypal anarchist. Ewing, hardly older than his graduate student, was tall and skinny and intense, and wore thick rimless glasses. He insisted on driving the truck alone, and set off optimistically for Norfolk, while the other two, in Crary's car, went by Wilmington for two hundred pounds of TNT. DuPont thoughtfully covered the outside of the car with signs reading "High Explosives." Ewing's henchmen were waved off bridge after bridge before, in a fit of civil disobedience, they tore off the signs and crossed the Delaware without hindrance. "Rutherford and I drove in turns," says Crary, "but between Richmond and Norfolk, about five in the morning, our second without sleep, I went to sleep at the wheel and we went into a deep ditch and turned over. When we got out, we looked back over a long stretch of straight road, and there were some headlights. We thought it must be Ewing. It was no longer dark, and as the lights came closer we saw it was his truck. We thought of course he'd see us and stop, and we practically were standing in front of him, but the truck didn't hesitate. We got a glimpse of him as he went by, hunched over, hanging onto that wheel, staring down the road. Five minutes later he was back. He said there began to be an idea in the back of his head that there was something on the road that had to do with him. We got to Norfolk finally, and we went to sleep, but Ewing was whisked off to Virginia Beach to meet the captain. He said he kept blacking out and waking up to find he was talking to someone and had no idea what he'd been saying."

"On the way back to Norfolk," Ewing said, "the captain had an accident. He was hospitalized, and his technician, who was to work with us, was killed. We sailed in some disarray. The executive officer took over, a literal-minded type, to whom the written orders were perfectly clear: we were not to interfere. We got little opportunity to work and came back most thoroughly discouraged. I took a train to Wilmington, where Professor Field met me. It was clear that we couldn't do anything on that ship. He suggested I do the land stations while he approached Woods Hole. Persuading them to carry explosives on their ships for the first time can't have been an easy job, but when we finished the land part we met him up there. We were regarded with some suspicion, but were told we could have a two-week cruise on *Atlantis* — a hundred-and-forty-six-foot steel sailing vessel that had been especially built for Woods Hole — after the regular season was completed. We did a few test shots right away on a little trial cruise near Cape Cod.

Doc and Joe in the Lehigh physics lab.

Atlantis, Floosey Belle and consignment for Ewing.

Vine brews a potent mix. Melting TNT.

They thought physicists were always seasick, so they sent along Columbus Iselin, who became director of Woods Hole during the war, in case we became useless. I suppose that my not having been seasick thus far — I had never been on the ocean — was largely from fury. Seasickness is like toothache, you know — you don't notice it if your house is burning. This was the chance of my life, as far as I was concerned. We sent Crary off in the whaleboat with a couple of sailors to make the shots. He took a pile of blasting gelatin and dropped the biggest charge eight miles out. The others went at regular intervals on the way back to *Atlantis*, where we had geophones on the bottom. Crary had a radio and signaled us at each shot. Columbus seemed a little surprised when the messages kept coming back clear, but when the whaleboat was close he looked and said 'He's seasick now.' I had to tell Columbus that Crary was just chewing tobacco. We had our brush with seasickness later. Rutherford was down below developing one of the records, with both arms in the sleeves of a portable darkroom; he suddenly turned green and said he was going to be sick, what should he do? I drew back my arm and said, 'I don't know, but if you ruin that record I'll break your jaw,' and he didn't have any more trouble."

During the thirties, young geophysicists melted TNT on the port deck of *Atlantis* (which was on loan for two weeks every summer from the Oceanographic Institution at Woods Hole) when the mates weren't looking, slipped in and out of the woods across the state of New Jersey, setting off explosions without permits, and drove home hungry after feeding Ewing's car, whose name was Floosey Belle, large rations of oil. Better fed and paid today, some are nostalgic. During term, field trips began after classes on Friday and ended at six or eight Monday morning; the intervening evenings were just sufficient for nursing equipment back to health. Other weekends were spent in the lab, and young Bill Ewing, age four or five, was a steady volunteer. "The campus police would open the gym at three or four in the morning so we could take showers after we finished in the lab," says Allyn Vine, who was a graduate student of Ewing's and is now a senior scientist at Woods Hole (whose research sub, *Alvin*, is named for him). "One semester when I was so broke that I lived in the lab, they'd come in if it was snowing and close the window and put a blanket over me. When we were ready to load up, we'd take the doors off Floosey and drive her right into the corridor of the physics lab. We weren't allowed the lab for our private use, but no one tried to stop us. Maurice was such a terribly intense guy. He'd work seventeen hours a day and go home to bed with journals under his arm. But he was never stuffy. He just always seemed to be right, though he'd let you argue to the last minute. Once he wanted to know if a piece of wiring was disconnected. I said it was; he thought not. Had I double-checked my positive answer? I was sure. He simply picked out one wire and cut it. There was a great flash of sparks, and all the lights went out. If he was thwarted, he was quite capable of being a mean, ornery guy. And so, I guess, were we. I remember one of the other professors was supposed to have asked, 'Why does Maurice surround himself with such a group of vicious young men?' Two weeks in the year was all the time we had at sea, and anything we got was precious. Everyone was welcome to make his own copy of any data, but God protect you if the original and one copy weren't in the file. He's said to be possessive about data like cores; well, Lamont's got the world's best core lab."

In any field, a few men make a mark by their own work, and a few make one through a generation of students they beget. Ewing appears to have done both in his need to learn about the earth. Of Lamont, someone has said: "Quite ordinary guys go there and come out leading the world." Just the number and distinction of the men in the earth sciences who were students of Ewing's would make him an interesting figure. Some are generalists like him, others specialists in any number of techniques for examining the earth. "The outstanding quality of any great teacher is the way he seems to challenge or welcome you into his circle," says John Brackett Hersey, a student in the thirties, director now of the Navy's Maury Center of Ocean Science, "and Ewing always had a very friendly, welcoming approach. He seemed to tell you, 'I'm glad you've come, I've been working on the most interesting thing, and I'd like to tell you about it, because I think you'd like it.'" John L. Worzel, a freshman in Ewing's physics class at Lehigh, was early relieved of a reflex camera, to the advancement of science and the study of gravity; Doc called him Joe, for reasons known best to himself, and Joe he is still. He found that while classes with Ewing were both easy and fun, classes with others were neither, and research with Ewing was not easy. Worzel developed a way of measuring gravity from surface ships fifteen years ago. He was Ewing's associate director at Lamont Observatory, and, he says: "I have felt I could do more good by freeing him to do something than if I did it myself." Two future geophysicists from Princeton, George P. Woollard and J. Tuzo Wilson, one now director of the Hawaii Institute of Geophysics, the other professor of geophysics at the University of Toronto, got Ewing to give them special seminars, for which they came over to Bethlehem. Woollard worked on seismic shooting at sea. About half the graduate students in Ewing's department, where they scarcely outnumbered the faculty, wanted to write a dissertation with him, a contemporary reports. J. B. Hersey was a Princeton man, but Ewing advised, a little, on his thesis. "When I came back from the oil fields to Princeton in 1938, Professor Richard Field told me to forget everything but where this fellow Ewing was, and become his student. So he was my thesis adviser; but immediately he went to sea with a leave of absence and a Guggenheim Fellowship, and the total advice I had of him was a half-hour meeting in the rotunda of Grand Central Station two years later, when we both were moving under orders from the Navy."

An unusual number of students have kept right on working with Ewing after getting degrees. For most, this has been a leisurely process, but (at least until he got a large endowment from the Henry L. and Grace Doherty Charitable Foundation a few years ago and Lamont became Lamont-Doherty), Ewing did not offer every inducement for working with him. "He always felt that if he could pay half the salary everyone else did, he could do twice the science," says an alumnus who finally left after ten years and one raise. "A lot of men have found it so stimulating to be in Doc's presence that they can't leave," explains a confirmed Lamonter, "while to others it's so nerve-racking they haven't stayed long." His teaching is more a way of being than a job. "He was the only teacher I ever had who fell asleep in his own class," says a geophysicist. "There were times when he came to class tired, badly prepared, obviously having worked all night," says Jack Oliver, professor of geology at Cornell, "and other times when he was brilliant, and would take you through a maze of math, theory, and wordy literature, making clear things you weren't sure even the people who wrote them understood."

"His strength lay not so much in giving polished lectures — he didn't do that — but in being a man who communicated by his demeanor and by doing it in your presence, first, the power of mathematics to solve basic problems in physics, and second, that basic problems were to be solved by people who worked hard enough," says Frank Press.

It has been common for successful researchers to use their research as leverage against having to teach, and Ewing's interest in students puzzled some of them. "Back in the fifties, when he was doing some of his most famous work," says Charles Drake, another student, now professor of geology at Dartmouth, "I asked him why he was a college professor, and he just looked kind of surprised and said, 'What else is there?' " Once, in the mess room of a ship five hundred miles south of the Cape of Good Hope, Ewing remarked, "One of the principal rewards of academic life is a body of former students you accumulate who are almost like the members of a family. I don't know any other profession where you find quite the same feeling of kinship."

Ewing had a fruitful, if irregular, liaison of almost twenty years with *Atlantis*, after Woods Hole loaned her to him in October, 1935. During their first venture, he got a good line of stations from Fortress Monroe to the edge of the continental shelf, and an accurate analysis of the shelf. Subsequent work corroborated and supplemented his records, and oil companies later found them intensely interesting. (Today people are going to court to keep oil companies off the Atlantic continental shelf, but in 1936, when Ewing hoped to extend his survey to other parts of the shelf, the chairman of New Jersey Standard Oil told him that he could not justify spending five cents of his shareholders' money on such a venture.) "By several masterly devices," wrote the geologist Walter Bucher, Ewing showed that the basement slopes down gently and evenly from its outcrop near Richmond to a greatest depth of twelve thousand feet beneath the surface of the sediments at the lip of the continental shelf, the end of the survey. The continental shelf is merely a thick wedge of sediment — the sort of place where oil forms — resting on the basement. Ewing wrote: "A seismic method was successfully adapted for making studies of marine geology. . . . Any new tool for giving information about the submarine portions of the earth's crust would be noteworthy because so few such are available." He was on a submarine making gravity measurements — another new tool which he was the first in this country to take to sea — before these observations were published.

Though the seismic work off Norfolk got what was rare at the time, precise and unambiguous information about the submarine portion of the earth's crust, it was the source of some aggravation of spirit to Ewing, because he had supposed it would mean something to someone. "I got those two fine men — Doctor Bowie and Professor Field — their numbers, thinking that then they'd know what to do with them. But they didn't seem to. It was only gradually afterward that I tried picking out their significance myself. I was then in transition from thinking of myself as a physicist and sort of technician for the geologists." Since he had found that the continental shelf is just mud, the most obvious significance was that the edge of the shelf was not the edge of the continent, and that the real edge — if indeed there was any difference between continent and ocean — had not been found yet and was somewhere still farther out to sea. Ewing took his gear out from a hundred to a thousand fathoms. "A thousand fathoms seems like a wading depth now," says J. L. Worzel, "but for the time it was comparable to NASA's problems today." A seismograph that went to the bottom required five hundred pounds of protective housing frugally contrived from the barrel of a naval gun. The logistics of it were burdensome to Floosey Belle's springs and Ewing's back. ("What did those two scallywags Vine and Worzel do but change ends, leaving me holding the middle?" he complained to a friend.) Explosions in the water create huge bubbles. After the direct waves of the explosion, the bubbles' sudden expansion and collapse make more sound waves, and some of each set go to the bottom and some to the surface, and some are reflected from the bottom to the surface and some from the surface to the bottom, and back again, and so on. The listening seismograph draws a mare's nest. Ewing put his explosives on the bottom, but there, at first, under enormous water pressure, they often failed to go off.

For one cruise he tried lowering charges and hydrophones on a cable, and also arranging them, three miles down, so that the hydrophones were in a position to receive, not be blown up. It was all very difficult — the bottom can be hard to find. The next year, each hydrophone (with its own seismograph) was ballasted separately and rose sometime after the shots. Ewing's deep-sea timed-release module was a bag of salt. He

Wrestling the deep-sea camera and its float over the rail.

learned a high value for time on *Atlantis*. His gear went into the water before dawn and might be recovered by evening. During his annual two weeks, he would get three or four good records. Other men, with their work, shared the ship. Steaming out to sea and back to Woods Hole consumed at least several days. If the sea was rough he would lose equipment. And in 1938 *Atlantis* had to run off before a hurricane, making thirteen knots under bare poles. "At sea, you're limited," says an oceanographer, "and if you're also a limited person, you're stuck."

While he waited for his seismic equipment to sink to the bottom, work, and bob up again, Ewing photographed the bottom. He built the first deep-sea camera, although he had found a notable lack of enthusiasm among experienced oceanographers for the idea. A camera had never been used in really deep water. Expert opinion insisted that the water near the bottom was so murky that pictures would show nothing. Ewing was discouraged from building a camera, and denied financial support when he went ahead. Preparing a camera was simple for men who put a seismograph on the bottom (though the first year someone put formaldehyde in the battery instead of distilled water). In 1940, Ewing, Vine and Worzel photographed current ripples in the white sands of the Georges Banks — familiar to generations of fishermen and navigators — and an abundance of animal life and tracks both there and south of Cape Cod. Bottom in the Gulf of Maine was dark and silty, with little life, changing to bottom of pebbles and cobbles toward Georges. The pictures were uniformly clear and sharp; the only difficulty, Ewing wrote wistfully, was "to find an interesting subject and to put the camera in focus with it . . . and to get the camera back afterwards." But ripples and bare cobbles meant currents, which more experienced oceanographers had "proved" did not exist at the bottom. Scientists were

Cold Texan with release and timing mechanism for the deep sea — a block of salt.

The difficulty was to find an interesting subject and to put the camera in focus with it. Sand ripples made, and eroded, by currents, southwest Pacific.

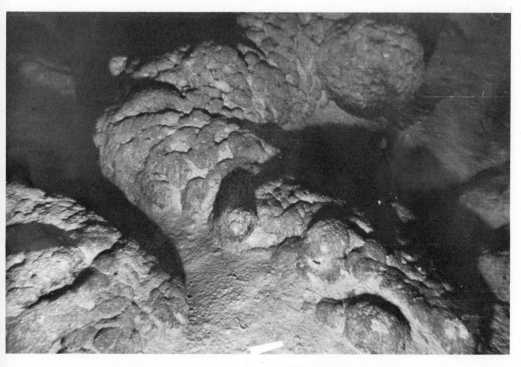

Pillow lavas on the flanks of the Mid-Ocean Ridge.

astounded by them, and pictures of animals and animal tracks were intensely interesting to biologists. The camera has since become a basic tool of oceanic research, and pictures have been taken in the deepest reaches of the oceans, over seven miles down, where they show animals, tracks, sediments, ripples, and rocks. The final prewar camera model was Ewingesque. A framework held the camera and held in its view: a compass (to orient the pictures), a pennant (to show any current, and its direction), a pendulum (to show the current's strength), and a white card (to show the relative murkiness of the water). There was even a glimpse of the bottom — of which a tube took a sample. It was lost in 1941 at fifteen hundred fathoms.

The first time Ewing's seismic gear was tried at a thousand fathoms, all the shots exploded and were received. It seemed phenomenal later on. But a tube in the recording mechanism burned out, and the echoes were not recorded. Ewing had gone to *Atlantis*'s hold to see if, with his ear against her plates, he could hear the shots going off. They were loud and distinct, and he listened for close to five minutes to the echoes that bounced off the surface to the bottom and back. There were eight round trips in all, each over seven miles. This was man's first experience of long-range sound transmission in the ocean. (Whales have used it for eons.) During the war, when he thought it might be handy, Ewing badgered the Navy for a test. The Navy's experts were sure that the keenest hydrophone could hear the largest bomb no more than a hundred and fifty miles away. A four-pound SOFAR — the inevitable acronym — bomb was heard twelve hundred miles away. More recently, a charge exploded off Brazil was heard south of Tasmania, more than ten thousand miles away — about as much unencumbered distance as there is in the oceans. The sound is transmitted, as through a speaking tube, in a layer of water (changes in temperature above and in pressure below it reflect sounds back in and prevent their dissipating), which has come to be called the SOFAR, or underwater sound, channel. In 1943, Ewing thought that SOFAR should be used to rescue fliers whose planes crashed and sank (listening stations on the shore could locate the source of a sound within a mile); but the air commanders, with military wisdom, decreed that four pounds of TNT were too dangerous to carry in military planes. Now the Navy has oceanwide networks of hydrophones to locate crashed missiles; and other networks track, and identify, ships everywhere in the ocean by the sounds of their propellers. The first recordings of whales singing were made a few years ago with Lamont's hydrophones in the underwater sound channel off Bermuda.

In 1939, Ewing was appointed chief scientist of the first major American voyage of research in some years, a year-long circuit of the Pacific to sail September 6. He wanted very badly to shoot seismic profiles out in an ocean basin far from any continent. His younger associates claimed that Roosevelt wanted someone for the Japanese to commit atrocities on; and when war was declared in Europe, and they were boarding their train, the expedition was canceled. In the fall of 1940, with his henchmen Vine and Worzel, Ewing took indefinite leave of absence from Lehigh, which had just promoted him from assistant professor of physics to associate professor of geology. Columbus Iselin had asked him to Woods Hole to do defense work. "I hadn't realized how important it was to Columbus, but he had never seen anyone before who could do math and physics and not be seasick," Ewing says. Besides having a valuable stomach, Ewing already could tell the Navy quite a lot about the sound in the ocean; and during the war the Navy improved electronic equipment like hydrophones so much that Ewing could do his seismic work without lowering

the gear all the way to the bottom. The government millions
that go to science these days are the product of the impression
scientists made during the war. Ewing and Iselin foresaw that
the government would find itself badly in need of services such
as theirs, but the need was apparent to the government only
after Ewing, Vine, and Worzel had worked at Woods Hole for
several months without salary, borrowing round robin in a
state Worzel indulgently calls "practically communism." By
January, they did some vital work on sonar. Though Iselin and
Woods Hole always got credit, it really was Ewing's work,
according to Captain Paul Hammond, a long-time associate
and trustee of the Oceanographic Institution. The thought of
submarines and war was already causing knowledgeable people the most extreme anxiety. Through a layer of water, the
very bottom can seem elusive; submarines were a horror.
Sonar is the remedy of choice, but in 1940 no one had much
experience with it. It had been around for nearly thirty years,
but was crudely developed and intermittently used. (The so-
called deep scattering layer of migrating plankton was discovered with sonar during the war; but that the layers rose to
the surface after dark was not discovered until after the war,
because the ships had not been out at night before.) Scientists
had encountered great administrative resistance to putting it
on naval vessels, and only a sprinkling of them had it. Echo
ranging was invented after the *Titanic* sank, as a means of
detecting icebergs up to two miles away in fog; the depth-
sounding and submarine-locating possibilities were suggested
several years later. For a few years, some lightships — *Nantucket Shoals* was one — were equipped to broadcast a sonar
pulse; ships with listening gear could receive the signal when
lights and even horns were muffled by fog. As a submarine
detector, however, sonar was confusing. Sometimes during
practice sessions it tracked subs with uncanny accuracy, and

sometimes the destroyers attacked imaginary submarines and steamed without pause over entirely real ones. "Some days you could see his periscope over there and not get an echo," said Ewing. After reading the Navy's records in Washington, he outlined a report, "Sound Transmission in Seawater," which, according to Worzel, who completed it by January of 1941, is still the Navy text on the subject. It pronounces that insubstantial objects return echoes. In the water, layers of different temperature are created by winds, waves, currents and the heat of the sun; and their boundaries, like the strata in the earth, reflect sound waves, or bend them so they never return to the ship. The Navy had been assuming that the ocean gets steadily colder with depth. It gets steadily colder with certain exceptions. To help the Navy find the exceptions, Ewing began improving a gadget called a bathythermograph. It had been invented by Athelstan Spilhaus for oceanographic research and made a continuous record of water temperature as it was lowered. Unfortunately, it took even longer to register one temperature than it did to pronounce; and old-fashioned Nansen bottles, lowered on cables, gave better data. By the end of January, Ewing, Vine, and Worzel — rescued from anticipated starvation by the first government contract for research in underwater sound (on which several hundreds of millions of dollars have been spent since) — had, streamlined eventually even to its name, a BT that registered temperatures in half a second instead of in over a minute. "One evening we designed something we thought would tow well and work quickly," said Ewing, "but we weren't smart enough then to patent it." In thirty years, he said, the efforts of many researchers, at unknown expense, have succeeded in improving the instrument by moving the hitch ring half an inch. Vine created a winch that would lower BTs at a constant rate from a moving ship, whatever the ship's speed — it did, too, if sloshed first with a bucket of seawater. Worzel plotted a graph from the first BT data, Iselin one from Nansen bottle data taken simultaneously, and were upset to see no resemblance. In fact, the Nansen bottle data were present in the BT data, but completely lost in the detail — an experience that was to be repeated every time an improved gadget was taken to sea. Through the war, Navy men came to Woods Hole for sonar and BT school. With the BT, Allied destroyers hunted subs, and Allied subs evaded hunters by hiding below sharp temperature changes. The instrument is still indispensable to navies and oceanographers.

Ewing was chief of research in physics — mainly marine acoustics — at Woods Hole. He had Friday meetings in Washington, from which he returned with projects — sometimes to be completed by Monday — like a small recording tide gauge a diver could leave on a beach and pick up later, and others that never got invented. He found a house at Woods Hole, on the narrow, often fogbound, street that runs along the harbor past the Oceanographic, and bought a sailboat that he optimistically put in the water each spring and rarely found time to take off its mooring. More young men came to work for him. His brother John, a nephew, and one of their high school mates, all geophysicists now, arrived from Tulia, Texas, probably an unequaled per capita contribution to science; Ewing gave the youngsters tutoring in physics.

Early in 1942, a camera was taken to New London to locate the Navy's own mines, against which it was helpless. Later in the year, the Navy summoned Ewing to North Carolina to try to identify by photography a wreck thought to be an enemy submarine. The task of identification had already strained the Navy's resources. A week of operations by a small fleet of ships with several divers had recovered no evidence. Ewing and his colleague Worzel were allowed an hour on site, and developed twenty-two pictures, nine showing the wreck and identifying it as a U-boat 352. In one picture a hip boot could be seen protruding from the wreckage foot first, and the commander of a Navy tug identified a lost grapnel. The Navy considered it vital to identify wrecks along the coast, and thereafter it identified many. In the fall, Ewing got an emergency call to assist a Coast Guard cutter and a converted yacht engaged in a prolonged depth-charge attack near Nantucket Shoals. They reported that their submerged target, presumably a submarine, "although heavily damaged and making an oil slick, is still moving." On Woods Hole's serviceable but far from dashing ship *Anton Dohrn*, Ewing and a hastily assembled crew (the chief scientist cooked while his associates were dishwasher and messmen) found "the two attackers following their prey closely, guns ready for the quick coup de grace when the target could no longer postpone surfacing." A marker buoy from *Anton Dohrn* showed that the target was not moving. Then photographs showed a ship's plating covered with seaweed and a number of small starfish. (It turned out to be the ancient wreck of a Nantucket lightship.) "The attack ended there. In breaking off, the cutter got up to full speed, spent all her remaining depth charges in a salvo at the spot marked by our buoy, then asked us to lead her to Gay Head Light, reporting that the last salvo had knocked out all her compasses. . . . The chief scientist dreamed of Cervantes on the way back to Woods Hole and the *Anton Dohrn* received a unit citation."

After the war, Ewing moved to Columbia University with some of his group from Woods Hole, where he remained a research associate. For several years he had three small rooms in a basement vacated by the Manhattan Project, into which came equipment, data, samples, and new crops of graduate students; seismographs operated under a hatch in the floor. "My assignment," Ewing said, "was to establish a program of

45

instruction in the geophysical study of the earth." He fulfilled
it by looking about the earth much as he had the Lehigh Valley, and making longer voyages. "During the war we used to
talk about what fun we were going to have afterwards when
we took all these instruments that were being developed and
started doing science with them," Ewing says. In the government's sprawling commitments he spied backing for geophysical ventures. He and J. B. Hersey, who was at Woods Hole, got
a contract to survey a hundred and twenty thousand square
miles of the Atlantic floor between Bermuda and South America. "I said as sure as you're a foot high they're going to have a
rocket range right there, and we'll be able to get some support." While Ewing resisted playing hooky at Morningside
Heights, Hersey did nine months of sea work — seismic profiles
and the first continuous deep echograms — on *Atlantis*, which
before the war had been laid up most of the year and hadn't
had an echo sounder that could reach deeper than the continental shelf. The new continuously recording echo sounder
revealed features on the sea floor that existing charts did not
even suggest. In 1946, the Navy opened its Office of Naval
Research, and for many years supplied the only governmental
support for basic science, from particle physics to vision to
mass behavior and other equally naval matters; ONR's first
contract went to Ewing and Worzel for SOFAR. Ewing also
got a contract for marine gravity measurements — there was
free time on Navy submarines — and for some years new
graduate students in geology were hustled off to look at the
Pacific through a periscope before they saw the inside of a
Columbia classroom. Others did exile in Bermuda working on
SOFAR and underwater sound. Albert Crary undertook to
look for a SOFAR channel in the atmosphere; ten years later
Lamont detected nuclear tests with an atmospheric wave recorder. A seismic survey of Bikini Atoll confirmed Charles

Darwin's theory that coral atolls grew up from ordinary volcanic islands while a large area of the Pacific ocean floor sank;
Sir Charles Lyell, the great English geologist, did a jig when
he heard the idea, but later geologists had been skeptical. The
survey found coral limestone fifteen thousand feet deep.

Three

Lifeless Depths, Living Fossils, Lost Continents

PROGRESS IN SCIENCE, like progress in religion, has as much consisted of the exposure of error as the revelation of truth. In 1521, in the Tuamotu Archipelago, between St. Paul and Los Tiburones islands, Magellan had two sounding lines spliced together and lowered. They ran out twelve hundred feet without reaching bottom. It is said Magellan concluded he was over the deepest part of the ocean (which is twelve thousand feet deep there and averages fourteen thousand). Not every explorer even tried as much. Seventy-five years ago, two distinguished oceanographers, Sir John Murray and Otto Krummel, proclaimed this the first scientific oceanographic research. If so, it was typical of much that came after. Science approached the depths gingerly.

Edward Forbes towered above other nineteenth-century English naturalists in the eyes of the young Thomas Henry Huxley, who was not a bad compounder of oceanic fallacies himself in later years. (Darwin, Huxley added with youthful charity, "might be anything if he had good health.") Forbes was professor of natural history at Edinburgh; and, as naturalist on Her Majesty's survey ship *Beacon* in the Mediterranean in 1841, he had dredged bottom to a greatest depth of thirteen hundred and eighty feet, finding living creatures all the way down. His curious conclusion was that life in the ocean peters out with increasing depth, disappearing completely by eighteen hundred feet, leaving an "azoic zone," as he called it. He wrote: "As we descend deeper and deeper . . . the inhabitants become more and more modified, and fewer and fewer, indicating our approach towards an abyss where life is extinguished or exhibits but a few sparks to mark its lingering presence." Animal life, to be sure, requires plant life for its nourishment, and plant life, sunlight; and sunlight penetrates scarcely more than a thousand feet beneath the surface. The evident cheerlessness of the abyss had impressed an earlier explorer, but not to Forbes's conclusion. During an expedition to the Arctic in 1818, Sir John Ross recovered worm-filled mud from six thousand feet, "thus proving [he wrote] there was animal life in the bed of the ocean notwithstanding the darkness, stillness, and immense pressure produced by more than a mile of superincumbent water." Disposing of Forbes's exceedingly reasonable fallacy occupied much of the energies of a generation of naturalists who followed him to sea; like a germ of the mind, it developed an immunity to medicine. As investigators pressed their searches farther than Forbes had his, they pushed the lower boundary of life deeper and deeper. Its defenders reluctantly allowed the azoic zone to retreat into the abyss, until it touched bottom.

In the 1850s and 1860s, the laying of transatlantic and other submarine telegraph cables increased interest in the ocean floor and pursuit of the azoic zone. On the survey ship *Bulldog*, sounding the Atlantic for a telegraph cable, there arose the ingenious idea of leaving the sounding lead on the bottom for an hour, so animals might cling to it. The lead returned from seventy-five hundred feet with thirteen starfish attached, and the ship's log received the exultant entry: "The deep has sent forth the long coveted message." The writer was unduly optimistic. Defenders of the azoic zone contended that dredges had not brought back animals from the depths at all, but had been contaminated on the way back by creatures that lived at moderate depths, above the azoic zone. When, around 1860, one of the first submarine telegraph cables broke, and the ends were brought to the surface for repair bearing a variety of living things, including a deep-sea form of coral that could only have grown on the cable while it was in place on the bottom, more than a mile down, the azoicists showed considerable agility in reversing themselves and insisting that the azoic zone was above the bottom (but still below eighteen hundred feet). After the rest of the ocean and its floor were fully populated by the taxonomic mind, it was still held that the greatest deeps, the trenches thirty thousand feet and deeper, might be devoid of life. The azoic zone lurked there for almost a century; when the *Galathea* expedition set out in 1950 there was still some thought of finding it there. Since 1950, fish, shrimp, and other creatures have been photographed repeatedly in the greatest depths that have been found.

In the summers of 1869 and 1870, Charles Wyville Thomson, the successor to Edward Forbes's chair in natural history at Edinburgh, made extensive dredgings in the North Sea from H.M.S. *Porcupine*, which not only helped exorcise his predecessor's azoic zone but recovered quite unexpectedly some most remarkable animals. One was a sea urchin unlike any to be found on the shores; its kind had flourished in the seas of the Devonian period, and were known only by their fossils. Thomson was conscious of looking at something thought dead for over a hundred million years: "We were somewhat surprised when it rolled out of the [dredge] bag uninjured, and our surprise increased, and was certainly in my case mingled with a certain amount of nervousness, when it settled down quietly in the form of a round red cake and began to pant — a line of conduct, to say the least of it, very unusual in its rigid, undemonstrative order. . . . I had to summon up some resolution before taking the weird little monster in my hand, and congratulating myself on the most interesting addition to my favorite family which had been made for many a day." The discovery caused the greatest excitement. For some years "living fossils," as they have been called, were the subject of intense popular interest. Hunters in Tennessee had reported hearing the screams of woolly mammoths from beyond just the next ridge. Fossil hunting was a popular occupation for amateurs as well as professionals, but the conclusion that hordes of species have been wiped off the face of the earth was not at all popular. Even to so progressive a geologist as Sir Charles Lyell, the intellectual odor of extinct species was revolting; naturalists were fervently seeking to explain all they saw by natural causes, recoiling from the fabulous catastrophes their forebears had evoked at every turning. *The Origin of Species* (*by Means of Natural Selection*, to complete the original title) described extinction as a natural process, but ten years later the idea of "living fossils" was scarcely less appealing than before.

Wyville Thomson's discovery stimulated the liveliest attention. Nature, red in tooth and claw on land, perhaps was milk and honey under the sea; the continental hinterlands having yielded forth no woolly mammoths, yet the depths of the ocean might hold great secrets. The Royal Society and the Royal Navy mounted an expedition with a ship, *Challenger*, to go around the world studying the depths under Wyville Thomson's direction; and the recovery of living fossils was one of the expectations. Wyville Thomson never accepted the theory of evolution, and as the *Challenger* voyage was outfitted, he wrote, of species: "The [vertical] range of the various groups in modern seas corresponds remarkably with their vertical range in ancient strata." The deeper you reach into the ocean, that is, the more ancient and primitive creatures you find. Thomas Henry Huxley and Louis Agassiz agreed. The idea sounds reasonable; Lamont Observatory, and other scientists, were still disproving it only fifteen years ago. Not the quenching of life, as Forbes had thought, but surcease from change lay in the depths; the ocean floor was a sort of rest home for unviable species, the ocean a protective wing under which might huddle the creatures who were not naturally selected. The eminent German biologist Ernst Haeckel went further than Thomson. Covering the bottom, he said, was the *ur schleim*, the primitive, undifferentiated protoplasm — the original life — from which all living, cellular organisms have descended. There was quite a stir when Thomas Henry Huxley found *ur schleim* in Edinburgh, in sample bottles remaining from the *Cyclops* expedition of 1857. Huxley, Darwin's bulldog — with his pugnacious Welsh face and stocky frame he looked the part — named the ultimate living fossil *Bathybius Haeckelii*; and the samples of other expeditions were searched for *ur schleim*. Bathybius was sublime. Bathybius, Wyville Thomson wrote, is

capable of movement, and there can be no doubt that it manifests the phenomena of a very simple form of life [it shows] no trace of differentiation of organs . . . an amorphous sheet of protein compound, irritable to a low degree and capable of assimilating food . . . a diffused formless protoplasm. The circumstance which gives its special interest to Bathybius is its enormous extent; whether it be continuous in one vast sheet, or broken up into circumscribed individual particles, it appears to extend over a large part of the bed of the ocean.

The German naturalists saw, with unconscious irony, "an infinite capacity for improvement in every conceivable direction." With Bathybius, nineteenth-century science seemed to have discovered the fount of life.

Challenger sailed in 1872. She returned in 1876. Although she found 4,417 new species and 715 new genera, the search for living fossils was unrewarded. The past was not at the bottom of the sea. Thomson continued to search eagerly for more living fossils even as the evidence mounted against them. "To the last," wrote Henry Mosely, the expedition's coral specialist, "every cuttlefish which came up in our deep-sea net was squeezed to see if it had a Belemnite's bone in its back, and Trilobites were eagerly looked out for, [but] as the same tedious animals kept appearing from the depths in all parts of the world, [our] ardor . . . abated somewhat. . . ." (Today some naturalists consider one of the *Challenger*'s best discoveries, an ancient little squid called *spirula*, with an internal shell, to be a belemnite; but Huxley, to whom the creature was entrusted, did not think so.) And Bathybius. After a decent interval, the chemist on *Challenger*, a Mr. Buchanan, clearly a mundane, unsympathetic man, examined Bathybius. Samples of anything in those days were commonly pickled in "spirits of wine." Mr. Buchanan found Bathybius could be created at will by adding brandy to seawater and letting them react. The *ur schleim* was the chemical precipitate, sulphate of lime, that resulted, a slimy substance that stuck to any particles in the water, giving a realistic imitation of ingestion. "Bathybius,"

Huxley wrote, "has not fulfilled the promise of its youth." And scientists are vulnerable to the hopes of their times.

This was the first great oceanographic voyage. It remains at least the longest. *Challenger* was a corvette of twenty-three hundred tons with auxiliary steam power; sixteen of her eighteen guns were removed to accommodate laboratories and scientific equipment including a hundred and forty-four miles of sounding rope and twelve and a half of sounding wire (the wire winch, of Lord Kelvin's devising, failed the first time it was used at sea). Her naval crew was commanded by Captain George Nares, the scientific party by Thomson. Her assignment was "the Conditions of the Deep Sea throughout all the Great Oceanic Basins." She sailed nearly seventy thousand miles in four years. A standard series of measurements, establishing a pattern for the subsequent expeditions of fifty years, was made throughout the voyage, standard and exhaustive. Steam was got up, sails were furled, and the ship was brought into the wind and held steady under engine as winches rumbled and lines bearing sinker, dredges, thermometers and water bottles went over the side. At three hundred and sixty stations, one every two hundred miles, staff and crew measured depth, temperature at various depths, atmospheric and meteorological conditions, direction and rate of current at the surface, and, occasionally, current at different depths. They dredged samples of the bottom and its plant and animal life, and samples of the water and its life from intermediate levels. It was, like all research, tedious work, and it rarely was finished before nightfall. Even the cabin boys turned out at first to see what the abyss might yield; but later there was less enthusiasm, particularly when, as it had an apparent fondness for doing, the dredge surfaced at dinner time. "Dredging was our *bête noire*," one of the officers wrote. "The romance of deep-water trawling and dredging in the *Challenger*, when re-

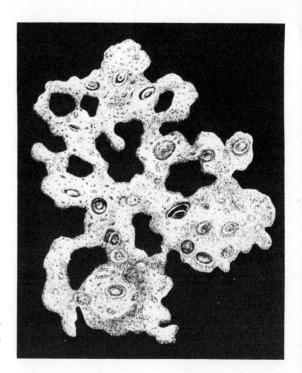

Bathybius, "mother of protoplasm," drawn by Haeckel from
microscope.

H.M.S. *Challenger* with dredge
and sounding gear ready to
lower into the abyss.

peated several hundred times, was regarded from two points of view; the one was the naval officer's, who had to stand for ten or twelve hours at a stretch carrying on the work . . . the other was the naturalist's . . . to whom some new worm, coral, or echinoderm is a joy forever, who retires to a comfortable cabin to describe with enthusiasm this new animal, which we, without much enthusiasm, and with much weariness of spirit, to the rumbling tune of the donkey engine only, had dragged up for him from the bottom of the sea." As for Robert, a battered parrot, from his perch on one of the wardroom hat-pegs, he cried repeatedly as the donkey engine rumbled: "Wha-a-t! two thousand fathoms and no bottom? Ah, Doctor Carpenter, F.R.S."

During the voyage, the scientists became able to forecast landfalls a hundred and fifty to two hundred miles away from changes in the bottom sediments. After the voyage, Sir John Murray studied the *Challenger* samples and named the different oozes — the official term — of the bottom. They are primarily the microscopic remains of planktonic surface creatures, shining opaline diatom shells, snowflakelike discoasters, delicately filigreed radiolarian skeletons, and, most common, the astonishing variety of homely snaillike shells of the tiny forams, particularly *globigerina*, after which foram ooze first was named. Murray classified his oozes by their origin — terrigenous or pelagic (far from land), the latter being either planktonic (from the surface) or benthonic (from the bottom). The planktonic remains are most numerous of all by far, and the oozes were chiefly interesting as libraries of surface water populations. Murray also named and explained the red clays, which are tan-colored and inorganic sediments covering the greatest depths, where the microscopic animal skeletons dissolve. He issued a map showing where in the oceans each ooze predominates, and he declared that only two kinds of sedimen-

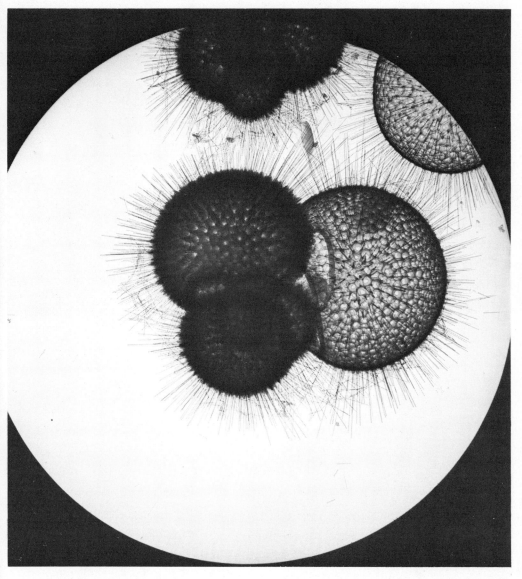

A foram through X-ray microscope. *Globigerinoides sacculifer* magnified
110 times.

Gothic filigree under the electron microscope. Radiolaria.

tation occur in the ocean — chemical precipitation and the slow accumulation of organic remains where they fall. (In the nineteen fifties a third kind was added, largely by the efforts of Maurice Ewing and others from Lamont Observatory.) Murray's classifications dominated the thinking of oceanographers for three-quarters of a century. Simply publishing the *Challenger*'s findings occupied the twenty years following her return; a large and international group of scholars was assembled — like those that gathered recently over moon rocks — to digest them into fifty large volumes, the last just in print at the end of the century. "Since the days of Columbus and Magellan there has been no such revelation of the surface of the planet," wrote Sir John Murray, who assumed direction of the *Challenger Reports* after Wyville Thomson died, his health undermined by the voyage. And Thomson — he had to get it all financed — had said: "Never has an expedition cost so little and produced such momentous results."

To sailors for centuries the ocean beyond the shallows of the continental shelf was, simply, "off soundings." Well into the nineteenth century there were many people, not only the unenlightened, who believed the ocean was bottomless. (It was also believed that wrecks would sink only to some unspecified depth and then hang there in a sort of limbo.) Scientists thought it reasonable to assume that the depth of the bottom below sea level would bear a relationship to the heights reached by the land above sea level. Early soundings seemed to suggest that the ocean *was* bottomless, at least to all practical men's intents and purposes. A Lieutenant Berryman of the American brig *Dolphin* reported a cast of lead and line in the Atlantic of thirty-nine thousand feet, and no bottom. A Captain Denham of H.M.S. *Herald* reported bottom at forty-six thousand feet in the South Atlantic, but a Lieutenant Parker of the frigate *Congress* ran out fifty thousand feet of line in the same area, with no bottom evident to him. "There are," Matthew Maury remarked patiently, "no such depths as these." Invalided out of active service in the U.S. Navy, Maury had turned to the study of the sea and become famous as a meteorologist and for the sailing directions he compiled, which greatly shortened travel along many trade routes (London to San Francisco by one hundred and eighty days, for example). He began trying to plumb off soundings in the eighteen forties. "Could it be more difficult to sound out the sea than to gauge the blue ether and fathom the vaults of the sky?" he wrote. (Maury was at the same time filled with wonder and with schemes for measuring its object.) "The most ingenious and beautiful contrivances for deep sea soundings were resorted to. By exploding petards or ringing bells in the deep sea,

"The surface of the sea [is] a nursery, its depths a cemetery for families of living creatures that outnumber the sands on the seashore for multitude." Once the bottom of a shallow sea, the Dover cliffs are composed of foram skeletons, each about the size of a grain of sugar.

when the winds were hushed and all was still, the echo or reverberation from the bottom might, it was held, be heard, and the depth determined from the rate at which sound travels through water. But though the concussion took place many feet below the surface, echo was silent, and no answer was received from the bottom." The experiment was performed in French Guiana in 1807, over a hundred years before the invention of sonar oscillators and transducers. Apparently no one listened with his ear under water, where echo may have been clamorous. Later investigators tried to fathom the sea by measuring variations in the force of gravity, and one to establish the mean depth of the Pacific from the speed and wavelengths of tidal waves. Maury himself devised a distance-measuring rotor, to be lowered on a sounding line, but no rope could sustain it. The first good sounding in great depth seems to have been made in 1840 by Sir James Ross, during a voyage with H.M.S. *Erebus* and *Terror* "to the utmost navigable limits of the Antarctic ocean." Sir James found bottom at 14,550 feet, "a depression, of the bed of the ocean beneath its surface," he observed respectfully, "very little short of the elevation of Mont Blanc above it." Sir John Murray called it the first good sounding; and in the 1960s a Navy ship went out and did it all over again with echo soundings, which were reported in *Nature*. It was gotten by old-fashioned line and sinker, but the fabulous numbers which line and sinker had been producing in deep ocean were avoided by unusual means. Two longboats were tied bow to stern. One held four miles of rope on a reel. The other held oarsmen, who rowed constantly, for six hours, against wind and current to keep the first boat in position. Such soundings were not made every day. Ross also timed the descent of the line, assuming, correctly, that he had reached bottom when the rate suddenly dropped. (In great depths the line itself is so heavy it continues to run out even after touching bottom.) Maury ultimately found he could achieve good results less laboriously with thin baling twine (marked every hundred feet) and a sixty-four-pound cannon ball as sinker, both twine and ball being disposable. He had had Navy ships required to make daily meteorological observations; now he got them stocked with sixty-four-pound balls and ten-thousand-fathom reels of twine. In 1855 he published a bathymetric chart of the Atlantic from 52° north to 10° south, incorporating the one hundred and eighty deep soundings that were then available. The first transatlantic telegraph cables were laid largely on the basis of his soundings.

With the invention of one of his midshipmen, Maury started sampling the bottom. "With this contrivance, specimens of the bottom have been brought up from depths of nearly four miles," he announced triumphantly. "The wonders of the sea are as marvelous as the glories of the heavens; and they proclaim, in songs divine, that they too are the work of holy fingers. Among the revelations which scientific research has lately made concerning the crust of our planet, none are more interesting to the student of nature, or more suggestive to the Christian philosopher, than those which relate to the bed and bottom of the ocean." The bottom samples from the North Atlantic teemed with the small remains of diatoms and forams; Maury sent them off to a Professor Bailey at West Point for professional description, but was moved to tell their story himself:

> Quiet reigns in the depths of the sea. . . . The unabraded appearance of the shells, and the almost total absence among them of any detritus from the sea or foreign matter, suggest most forcibly the idea of perfect repose at the bottom of the sea. Indeed, those soundings suggest the idea that the sea, like the snow-cloud with its flakes in a calm, is always letting fall upon its bed showers of these microscopic shells. . . . This

process, continuing for ages, has covered the depths of the ocean as with a mantle, consisting of organisms as delicate as the macled frost, and as light in water as down in air. . . . In the deep sea there are no abrading processes at work; neither frosts nor rains are felt there, and the force of gravitation there is so paralyzed that it cannot use half its power, as on dry land, in tearing the overhanging rock from the precipice and casting it down in the valley below . . . so long as the ocean remains in its basin, so long must the deep furrows and strong contrast in the solid crust below stand out bold, ragged, and grandly. Nothing can fill up the hollows there; no agent now at work, that we know of, can descend its depths and level off the floors of the sea. But it now seems we forgot the myriads of animalculae that make the surface of the sea sparkle and glow with life; they are secreting from its surface solid matter for the very purpose of filling up those cavities below. Those little marine insects build their habitations at the surface, and when they die, their remains, in vast multitudes, sink down and settle on the bottom. They are the atoms of which mountains are formed — plains spread out.

This is a picture of the ocean floor that prevailed for a hundred years. *Challenger* only added detail, and *Challenger* dominated oceanic research until World War II. Where Maury sang, Murray counted, classified, and documented. Nothing of comparable novelty or importance was found in the oceans — certainly not on the bottom — for seventy more years. After *Challenger*, the published results of marine biology exceed the combined results of all other oceanic research. The *Challenger Reports* "remained the main source of knowledge of the ocean floor until the nineteen thirties," writes Sir Edward Bullard. "In retrospect it is odd how little was done. . . . The oceans were left largely to the biologists." The biologists were very busy. Alexander Agassiz concluded — on zoological grounds — that the isthmus of Central America is geologically recent, and that the Atlantic and Pacific once mingled freely in the tropics. (Agassiz was a tireless outfitter of expeditions, though he suffered all his life from seasickness, which he reported was somewhat assuaged by an exclusive diet of pineapple, fried bananas and sauerkraut.)

Near the turn of the century, Sir John Murray wrote, with the authority of thirty or forty years' study: "There is no evidence that any portion of the present land ever lay under the deeper parts of the ocean. Nor, on the other hand, can it be shown any part of the existing ocean abysses ever rose above sea-level into dry land." But no one listened to Sir John Haeckel, the biologist who covered the abyss with *ur schleim*, also choked the Indian Ocean with a "lost continent" of Lemuria, which he contended was required to account for the distribution of lemurs on the adjacent continents. Into the western North Atlantic, American geologists tucked a lost continent of Appalachia, which, they said, was required to supply the sediments that the like-named mountains are constructed of. Europeans salted away a continent in their part of the ocean. The oceanographers pronounced the redundant ocean floor featureless as well (a description better suited to the state of knowledge). "Being protected by water from rapid sub-aerial erosion, which sharpens the features of the land, and subject to regular accumulations of deposits, the whole ocean floor has assumed some approach to uniformity," wrote Otto Krummel, echoing Maury. The geologist Archibald Geikie agreed, adding: "The only motion possible is the slow creeping of the polar water along the bottom. . . ." Agassiz was vehement. He wrote: "The deep sea floor is a monotony, only relieved by the dead carcasses of animals which find their way from the surface to the bottom and which supply the principal food for the scanty fauna found there." *The Oceans*, a textbook published in 1942 and still the most comprehensive, remarks: "From the oceanographic point of view the chief interest of the topography of the sea floor is that it forms the lower and lateral boundaries of the water."

In the nineteen twenties, the nonmagnetic ship *Carnegie* measured the earth's field at sea until she burned in the harbor of Apai, Samoa, in 1929; the Dutch physicist F. A. Vening Meinesz developed a method of measuring gravity at sea from submarines, and used it at several hundred stations in the Dutch East Indies with results that were both fascinating and ambiguous; and the German ship *Meteor*, commissioned in part to assess the possibility of reclaiming gold from seawater for payment of wartime reparations, studied the circulation of water in the Atlantic. Three-quarters of a century after Maury there were fifteen thousand deep soundings in all the oceans — a meager one per thousand square miles, had they been scattered evenly — to which the *Meteor* and others added a few sonar tracks before World War II. In January of 1934, for example, Sunday science pages breathlessly reported the just-completed voyage of Captain Claude Banks Mayo, USN, from Hawaii to Japan, and 17,230 echo soundings that were interpreted, with more enthusiasm than scrutiny, as showing the mountains, valleys, former river beds, and plateaus of one of those lost continents. A new era of exploration began in the late forties with instruments largely developed or perfected during the war and with money from newly interested navies. Depth, magnetic, gravity and other measurements were ultimately made in sufficient numbers to begin illuminating the ocean floor, which lost both its featurelessness and its sunken continents, as earlier it had its *ur schleim*.

Four

A Scene the Most Rugged, Grand, and Imposing

In the summer of 1947, Maurice Ewing led his first large expedition; he had gotten *Atlantis* all to himself for two months, about as much time at sea as he had had in all the years before the war. "We found a large percentage of all the things we've worked on until today," he once said. His voyage was noteworthy because he made the first examination of an ocean basin with a number of modern instruments; also, Ewing was not an oceanographer, or a marine biologist, but a geophysicist. Nominally commissioned, by the National Geographic Society, to explore the Mid-Atlantic Ridge (the most important feature in the ocean, as he showed later, but then the only significant one known), Ewing lost no opportunity to scrutinize anything that lay between Cape Cod and the Azores,

The *Atlantis*.

where *Atlantis* refueled. To the men who went, the cruise, called *Atlantis*-150, was to be a scientific event of the first importance, applying to a vast area of the unknown in one burst all the accumulated resources of contemporary technology as had that first great oceanographic voyage seventy-five years before. If, as Sir John Murray wrote of *Challenger*, there had been no such revelation of the surface of the planet since the days of Columbus and Magellan, then, says a scientist at Lamont Observatory, "those of us who went out on *Atlantis* did so in the spirit of the *Challenger*." There was a sense of high adventure on *Atlantis*, and no one was disappointed.

The lay of the land ahead was unknown except in grossest outline. Beyond the continental shelf — off soundings — the topography of charts was almost entirely fictitious; believing the sea floor uniform, cartographers drew contours around soundings tens and even hundreds of miles apart. Even echo soundings — until the year before when Ewing got *Atlantis* a machine that recorded continuously — had been made spasmodically when they were made at all; and the best machines were not intended to penetrate below 2,000 fathoms — 12,000 feet — a depth that is reached one hundred miles from Cape Cod. Lack of knowledge ably supported the prevailing view that the ocean bottom was featureless. One feature was too large to be missed. In the 1850s Maury published a few shallow soundings from an area in mid-ocean which came to be called Telegraph Plateau. But, curiously enough (such was the difficulty of sounding and charting the ocean floor before sonar), it was the temperature of bottom water on each side of the Atlantic, measured during the *Challenger* expedition, which demonstrated the existence of a continuous ridge of high ground running almost from pole to pole, dividing the water into two basins. The Mid-Atlantic Ridge rises from beneath the reach of the early echo sounders to a depth of around a thousand fathoms. If its presence is too obvious to be missed, its nature was not. Ridge, mountain range or ranges, and high plateau all were proposed. One geologist believed it the first arching up of the earth's crust in an incipient continent, while another authority had it the debris left when the continents split apart and moved to their present positions — nor were devotees of a lost *Atlantis* silent. In the mid-forties, foremost geologists spoke ambiguously of a Mid-Atlantic Swell. "The vagueness of most sounding data, and the lack of complete geophysical information, have resulted in numerous and sometimes wondrous hypotheses concerning the history of the Atlantic," Maurice Ewing and Ivan Tolstoy wrote in the *Journal of Geophysical Research* in 1949, reporting some of the findings made from *Atlantis* in 1947. The efforts of previous expeditions, however painstaking, being almost entirely worthless geologically, much remained to be discovered, and everything to be realistically described.

Atlantis sailed from Woods Hole in mid-July, with a scientific party of eight. With just the recording echo sounder (which Ewing and Worzel went at with screwdriver and soldering iron until it recorded to four thousand fathoms instead of to two thousand), *Atlantis* was incomparably more powerfully equipped than any earlier expedition to understand the Ridge or any other feature that might be found on the ocean floor. But Ewing also had stocked her with every conceivable kind of gear for examining the earth's surface at sea — except magnetic gear, which he carried the next year, and gravity gear, which required submarines. He even had Nansen bottles to sample the water, plankton nets to sample what was in it, and BTs to take the water's temperature. In the absence of Woods Hole's expert on sediments, Henry Stetson, he borrowed the Stetson coring device, a long tube surmounted by much lead, a cookie cutter — as Ewing liked calling it — that took out of the bottom a mud cookie one and a half inches through and eight or ten feet long. He proposed to take some cores along the way and, since cores weren't his thing, to bring them back to Stetson. Columbus Iselin thought it was a generous gesture. Ewing remarked ingenuously: "I felt a sense of obligation, with an expedition entrusted to me for two months, to get data of every conceivable kind." Unexpectedly, there came along with Stetson's corer his assistant, a young micro-paleontologist named David Ericson. "I don't think anyone asked me to go," says Ericson, a soft-spoken, literate man with a full, sea-captain fringe of graying whiskers, who became Lamont's expert on Pleistocene sediments and climates. "I just stepped aboard and no one questioned me because it seemed the natural thing to do." To anyone who didn't know Ewing, the cruise came as a bit of a surprise. "When you went out with him," says Ericson, "There was no English or Swedish social life in the main cabin in the evening. You were working night and day. Every minute you could see he was thinking of the data he was getting and the money he was spending. He was a man of extraordinary enthusiasm and just wrapped up in what he was doing." It is difficult to even imagine him going to sea, as another scientist does, with a flexivan fitted out with all his instruments as well as a settee, air conditioning and a bar.

In a few days, *Atlantis* was in tropical waters; after a brush with bad weather and a brief call at Bermuda she headed into mid-ocean. The water had become very blue. A few flying fish landed on the deck at night, offering themselves for breakfast, and were collected by the watch. Some men swam during stops for hydrographic stations, protected in one of the big trawl nets from sharks. Every day there were new experiences to understand. "Ewing was always the natural scientist," says Frank Press, a student then, "interested in everything, the St.

Elmo's fire at the top of the mast, the nature of swell, the flash at sunset." As they sailed with echo sounder continuously recording, Ewing said, "I had the strange feeling of being an aviator flying high over an unknown planet tantalyzingly hidden from view but outlined on his radar screen." It was a picture that a few men had tried to imagine, and some would have given much to see. For a hundred years men who were curious about the land hidden under the oceans had floundered in misleading analogies and comparisons and inferences because there was no way of getting more than the slightest trickle of real information. Suddenly there was a flood.

Atlantis was a day out of Bermuda, in three miles of water, when the bottom began to shoal gradually. As she sailed on, it rose beneath her for hours, until after twenty miles she was over the flat top of a small seamount two miles high. Ewing decided to take his first core. "I told Press, 'Here's our chance to get a core without gambling a lot of time.' I was so jealous of ship's time I was determined I was going to take a core in shallow water to gain experience. Well, some people, you know, never get over their first anything, their first drink, their first piece of tail. As it happened that core was one of the best of my life." The coring tube with its half ton of lead weights was swung overboard, and cable began cracking across the deck as the winch in the hold turned and the coring device dropped out of sight. Bottom was at eight hundred and forty-one fathoms. In a quarter of an hour the winch was slowed to a creep. Ewing peered at a tension meter on the cable; there would be the slightest tremor at the instant a trigger let the corer fall into the mud. The weight of the mile of wire was greater than the half ton of lead; it would not go slack. Too much cable let out would kink and break later under strain, nor was too little desirable. Ewing waved, the winch was reversed, and the cable came back, taut, snapping onto the winch drum with reports like gunfire. The core was full of surprises. "In the old days," says David Ericson, "the theory was you could take a core anywhere at sea and get a complete history of the world from its creation — if you could take one that was long enough. There would be fossils of the first animals of course; and we expected, all of us on that *Atlantis*-150 cruise, that maybe even a piece might be sticking up of the original crust of the earth that had frozen out of the molten state." Oceanographers and paleontologists regarded the ocean sediments romantically and simplistically as an "epic poem of earth history," as one paleontologist later wrote, ruefully. "The assemblage in the tops of cores could be used to determine the areal distribution of various species at the present day, while samples from lower levels could show, in an orderly manner, the development of present day species. . . ." The sediments accumulated by what was repeatedly called an endless snowfall or a steady rain, of tiny planktonic shells and skeletons from the surface. The bracing moral lesson was that everything could be learned by sufficiently close examination of the evidence — that what the snowfall put in, the patient investigator could, particle by particle, extract as information.

"I spoke then," Ewing says, "of finding such events as the first plant life and the resulting change in the atmosphere, and the appearance of the first carbonate-shelled animals. We had every reason to believe the sediment column represented the whole of geologic time."

"Somehow," says Ericson, a little wistfully, "we never seemed to find those trilobites everyone was talking about getting. When we examined that first core we found Recent sediment on top, and Eocene sediment at the bottom. That in itself was exciting, because in fact, sediment older than a few thousand years had never been recovered in the ocean before. Actually it had never been recognized; as soon as we got back to

Woods Hole, I looked at some cores that had been minutely described as Recent, and found they were in fact much older. The Eocene lasted twenty million years — mammals became pre-eminent — and ended forty million years ago. Reaching it in our first core did give us hope of soon getting sediment that was very much older. But when we examined the core more closely we found something rather more unexpected. The eight inches of Recent sediment lay directly upon the Eocene. Of the intervening forty million years there was no trace. It was scarcely an example of the complete record that orthodox theories had led us to expect. What had happened to it? On that point the core itself made no sense; yet all accepted ideas led one to believe a core could be understood completely by minute enough analysis. The Recent sediment was coarse, and had undoubtedly been laid down on the top of a mountain where any fine stuff could be winnowed away by currents and lost; by the same argument, the Eocene sediment was not laid down on a mountaintop — it was extremely finegrained. Both sediments were soft and as fresh as the day they were laid down. One was led to believe the mountain was raised in the interim, and the missing sediment shaken off, if one could believe the violence of making a mountain could leave the Eocene sediment unaffected, as no one could. We had to leave the question, with the suspicion, which grew as we went on, that the ocean was more than oceanographers accounted for."

Whenever anyone was not working or sleeping there were discussions of the new things they were seeing. "But Ewing never forgot that the ship was expensive," says Frank Press, "and we worked it twenty-four hours a day. A cruise with him was brutally hard work. I used to bunk with him, and even when he went to sleep it was only for an hour. There was nothing he wouldn't do himself; if a hundred pounds of TNT

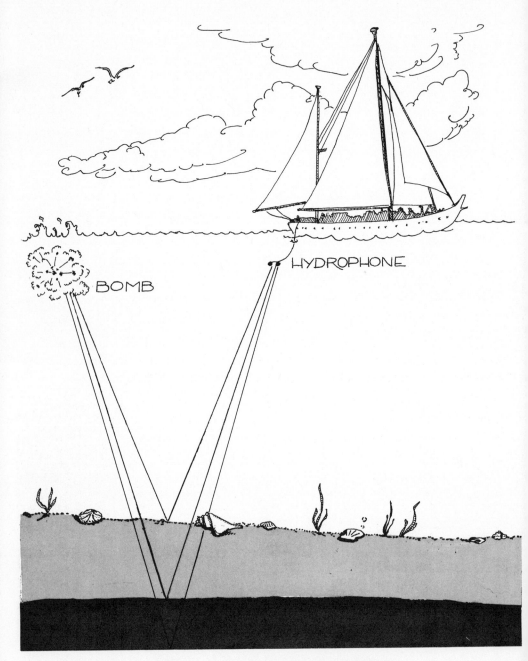

Explosion wave paths during seismic reflection.

Seismic shooting went on at half-hour intervals.

needed moving, he moved it. The cruise had no elaborate plan, we made it up as we saw what the bottom was. We worked up our data at sea in those days. We knew what we had. As soon as the seismic records were out of the photographic darkroom we had them analyzed — all day, all night, every day — and so called the next shot by the last one. The sediments that they showed were astonishingly thin everywhere we went, nowhere as deep as a mile and in some places too thin to see on the record." Seismic shooting went on as fast as human beings could do it, drawing an intermittent profile of the earth below.

"I remember one used to hold the fuse in one's teeth," says Ericson, "sitting on the afterdeck. The charge — a half pound of TNT — was in one hand and a lighter in the other. One wasn't supposed to put the fuse in until the last minute. Then you flung it. This continued at half-hour intervals day and night. He would seize every opportunity to get information. There would be two-day stretches when he was going continuously. If the topography became surprising the ship would circle around and try to outline it. With him you were either going to sink the ship with too much explosive or discover something interesting."

"I was desperately interested in shooting seismic reflection profiles in the deep ocean basins."

"I recall one time a charge did go off against the side of the ship. Down below — where the shots ordinarily sounded like someone hitting the hull with a sledgehammer — there was a *tremendous* noise. We were sure this was finally it. But all that happened was one hull plate was slightly bent in."

Ewing had been trying to learn more in less time. In the thirties, he had put all his seismic gear on the bottom, and got at best one or two measurements for a day's work. After the war, with instruments developed for the Navy, he worked from the surface. The ship was stopped, hydrophones were lowered into the water, and a charge of TNT was thrown over. Echoes came back from the earth below and the deeper strata; when they were recorded, the hydrophones were retrieved and the ship got under way again. Ewing wouldn't even sail *Atlantis*, though he liked to sail, if the engine would get him on faster to more new places; and he now begrudged losing the better part of an hour on each measurement when the sound of an explosion got to the bottom and back in just a few seconds. "We said let's see if we can do it without stopping the ship," Ewing said. "We perfected underway seismic shooting during that cruise, a highlight of my life." Ewing stopped the hydrophones, instead of the ship. Just before a charge exploded, a towline was let free and the hydrophones floated still in the water until the echoes were recorded (if the hydrophones were moving, the sound of running water hid the echoes); then they were hauled back aboard. Ewing admitted it was a little hard on backs. But the records were good. Shooting stopped only when the ship was stopped for sampling, or when the sea was so rough she could not get away from the TNT charges. A three-hundred-foot line was tied to the explosives so they would tow near the surface, but not too near the ship. One memorable night the line fouled. The charge hung out of sight a few feet below the rail. Ewing, who was on watch, had had the habit of putting his ear to the rail since he first heard wave returns ten years before in *Atlantis*'s hold; but this time he used the rail on the opposite side of the ship. It was his first accident with explosives. For the next twenty years, until TNT was supplanted by a pneumatic gun, an essential part of the equipage of any Lamont vessel was a large supply of toy balloons to float the TNT — and as the ships steamed through the most distant parts of the oceans, scientists aboard them could be seen in the stern blowing up brightly colored balloons, tying them to chunks of TNT, and watching them float away astern, fuses burning.

As soon as a new record could be developed in the darkroom, and interpreted, it was time for another shot. "There was an air of imminent discovery," says Frank Press. Eleven days out of Woods Hole, four out of Bermuda, on July 28, around midnight, *Atlantis* sailed out of irregular, generally rough topography and onto — over — a plain flatter than anything anyone had seen before. The echo sounder showed twenty-nine hundred fathoms — seventeen thousand four hundred feet. For two and a half days, *Atlantis* sailed on to mid-ocean, and the plain below remained at twenty-nine hundred fathoms. The seismic reflection shots, which at that depth were accurate within a foot or two, showed twenty-nine hundred fathoms even when the echo sounder wandered off, as, for reasons of its own, it often did. The plain caused some amazement, not only on *Atlantis* but later among geologists and oceanographers ashore. It was so very flat. No known geologic process produced quite that appearance; the floodplain of a river may have the flatness, but has certain obvious dissimilarities. The abyssal plains, as they were to be called, are the flattest places on earth, and they are immense. In scale they are flatter even than the best billiard table. Sprawling, rough-edged, and shapeless, they are hundreds, and occasionally over a thousand, miles across. *Atlantis* found several more in the next few years on both sides of the Atlantic. They are in every ocean, but are most extensive in the Atlantic, where they nearly cover the basins between the Ridge and the base of the continents. The Pacific is ringed by trenches at the base of the continents and the island arcs, like Japan and the Philippines, and has few abyssal plains to seaward; but the bottoms of many trenches are narrow abyssal plains. The Swedish *Albatross* expedition found abyssal plains in the Indian Ocean in 1948; they concluded, because their coring apparatus came back bent and empty, that the plains were vast lava flows. Abyssal plains have been minutely described and their nature is axiomatic today, but in the forties it was subject to argument and doubt. Even in 1959, a Lamont paper noted, more sorrowfully than indignantly, a few scientists persisted in the lava flow theory. Ewing, although he put forward no theory of his own for several years, thought it nonsense to suppose lava could flow so far and flat in cold water; and he succeeded in coring the plain, getting sediment, not lava. The lavaphiles stuck to their lava; there might just be some sediment on top of it, they said, destroying their original argument. The extremity of the plains' flatness was concealed by the early echo sounders, which, at depths like twenty-nine hundred fathoms, were accurate to no better than twenty-five or fifty, and sometimes only one hundred, fathoms. They ran off regular ship's current, which itself was no thing of precision — when the ship's cook opened his refrigerator the echo sounder discovered a canyon in the ocean bottom. It was not always possible to spot the artifacts immediately; an early and prominent discovery on the Atlantic Ocean floor was later rescinded and privately named Sixty-Cycle Ridge. The records showed humps, bumps, and jogs in the abyssal plains which the advocates of lava flows made much of. Ewing believed his seismic profiles, which were notably accurate, and which, though intermittent during *Atlantis*-150, never varied from twenty-nine hundred fathoms for three days on the abyssal plain. Only after 1953, when Lamont built its precision depth recorder, which is accurate to within six feet in eighteen thousand, was the flatness of the abyssal plains demonstrated in a way that was clearly visible and hard to deny or ignore. A PDR record of an abyssal plain is striking, but boring — a straight line. The discovery of abyssal plains was one of the most important made in the ocean. They are an essential part in a complex chain of geologic phenomena, some of which, as obscurely related as certain oil strata, vanishing marine telegraph cables, and sand, had puzzled geologists for years. There were new puzzles on top of the old puzzles. For instance, the seismic profiles of the abyssal plains showed reflections within the sediment. Seismic reflections had always been translated as layers and strata, but what were these layers and how did they get there? The epic poem of the sediments was not thought to have been printed in installments. The rain of planktonic corpses was steady, the snowfall endless — just the appearance that sediments had everywhere but on the abyssal plain. Therefore some geologists were to insist the abyssal plain could not be layered either, except by some unidentified failure in the seismic process (which only abyssal plains could cause).

Just before midafternoon on July 30, the third day of crossing the abyssal plain, the echo sounder recorded a gentle rise of a hundred fathoms. In twenty minutes, depth was back to twenty-nine hundred fathoms. Two hours later there was another small hill in the plain; and from there the ground grew gradually and steadily rougher and higher until, almost exactly in mid-ocean — 1,650 miles from home and 1,680 from the African coast — *Atlantis* lay above the central highland of the Mid-Atlantic Ridge. The fathometer drew a wild, jagged country of unweathered pinnacles and abrupt chasms. There had been deep sediment on the flanks of the Ridge, but there was little or none on the crest. It was quite clear immediately that the Ridge was not a vague swell under a thick blanket of sediment; for a while there was some anxiety that it might be a meaningless jumble of rough topography. There were high mountains and deep valleys in profusion. After nearly a week of survey, *Atlantis* lingered over a particularly spectacular piece of territory. From a deep cleft, a mountain rose with exceptional steepness nearly ten thousand feet, a bold contrast that later *Atlantis* traced north over fifty miles. First, however, she stopped for samples. A heavy sea was running. Two coring devices were lowered to the bottom, one after the other, but the gyrations of the ship were so great, and the tensions on the cable so various, that both were hoisted again without having struck bottom, empty and still cocked. At 2 A.M. the heavier corer was sent down again; it reached bottom, stuck, and came free only as the strain on the half-inch steel cable approached breaking point. A rusty strand snapped and began to unravel. Ewing saw it — the strain on the cable was fifteen or sixteen tons — and, he said, "I thought, the place it breaks is the safest place to be. I had a roll of tape in my pocket and I walked over and began taping the cable. It didn't break." The tube probably had caught between rocks; it came aboard badly bent and containing only one small piece of freshly broken igneous rock. The small corer was completely wrecked in an attempt to sample the top of the mountain. Only the rock dredge remained usable and it was lowered — though getting a dredge on the bottom, and towing it on steep rocky slopes, are a good deal harder than coring. An earlier dredging had returned nothing. The dredge surfaced this time, after four hours beneath the surface, with a hundred pounds of igneous rocks of several varieties — olivine, asbestos, an altered serpentine — some much distorted by heat and pressure. The dredge promptly was relowered, to sample the floor of the cleft, and came back with four hundred pounds of basalt, including shiny pieces of pillow lavas, which form when hot lava cools quickly in water. After twenty-four hours on station, everyone, except Ewing, admitted to being tired enough to leave. *Atlantis* got under way and followed the topography north by echo sounder. Already more was known about the nature of the Ridge than had been discovered during the previous ninety years. Just the obviously mountainous nature of the Ridge was enough to scuttle most theories. That there were, on the Ridge's steep slopes, rocks to be had for the trying, was both stimulating and against all predictions. (It almost seemed that the truth of what lay under the ocean could be found merely by taking scientific beliefs and reversing them.) The dredge had been brought along to get sediment. Not only was the bottom of the ocean supposed to be covered, its shape muffled, by a thick blanket of sediment, but under the sediment the rocks were supposed to be old, original and undeformed. "When I was getting started in this business, we thought if you could look you would see the original crust of the earth," says a geophysicist.

Many of the rocks that came up in the dredge were fresh-looking, bare and unoxidized. The steep slope and deep ditch suggested to Ewing a fault scarp, where in relatively recent times a piece of the earth's crust had slipped up or down against another, with attendant earthquakes. A later dredging of metamorphosed, faulted, and shattered igneous rocks, and of sediment full of pulverized rock, lent some support to the idea. (The dredge hauls were reported after the cruise, and were also used in a series of papers on the composition of the Ridge — written in 1969.) During the cruise, echo sounder profiles were regularly laid out on charts along the ship's course and the topography roughly sketched in. As the voyage progressed some order was found in the Ridge topography, and with study more emerged during the winter. Abyssal plain, at depths of over three miles, flanks the Ridge. A region of low hills emerges from the plain, and gives way in turn to several levels of ever higher and more rugged ground. The backbone of the Ridge is a series of north-south parallel ranges, very jagged, with no flat stretches, rising from more than two to less than one mile below the surface — or two miles above the abyssal plain. Jagged, unweathered, basaltic, recently fractured and erupted, the local description of the Ridge is little changed today, except by greater detail, and is valid along its length. But it has become apparent that the Ridge is but part of the most important geological feature in the world.

Atlantis followed the Ridge north from the horse latitudes to the Azores. At San Miguel her tanks were filled with bunker oil, diesel being unavailable. She sailed pouring black soot from her stack; decks, sails, and men rapidly became disagreeably grimy. She sailed the Ridge another week and turned home. Then Ewing allowed himself the only seismic refraction profile of the voyage. *Atlantis* hove to to record the shots; her radar kept track of the distance to the whaleboat, which was dispatched with TNT, radiotelephone, radar reflector, and a sail. "They had a huge pile of charges in the bottom of the boat, and they sailed off way out of sight," says David Ericson, who was still accustoming himself to Ewing's seismic habits. The records were informative about the sediments, and implied surprising qualities in the basement rocks, below; they also showed that only with another ship, and a huger pile of TNT, would they be unequivocal.

Five

Torrey Cliff

"IF WE HAD FOUND the expected things, even Ewing might have lost interest, but instead it was like throwing gasoline on a fire," says David Ericson. "Almost nothing we'd seen had been as expected — the general lack of sediment in the seismic profiles, the youthful appearance of the Ridge, the wholly mysterious abyssal plain, and the several very puzzling aspects of the cores. For me, what he had found was so exciting that there was no question of whether or not I should work on the cores, simply whether I should do it at Woods Hole or Columbia. Of course, for developing a climatic record of the Pleistocene era, which I had my heart set on, those first cores were unpromising. What one looks for to get a climatic record is correlation: if one core shows the beginning of an ice age you expect to find it also in another core taken a hundred miles away. And in that first suite of cores the correlation was miserable. We were misled by layers of gray clay we thought might represent climate changes but realized later had quite a different cause; when we analyzed the fossils in the layers we didn't get any sense in terms of climate. So we realized that conditions in the ocean at any one time were not uniform — as had been thought — and we added the layers of gray clay to those puzzling and exciting sands and gaps of millions of years that we already know to be in the cores but didn't yet understand.

"Ordinarily, everything would have stopped there and we would have spent years studying the cores in Schermerhorn Hall. In the tradition of voyages ever since *Challenger*. But almost immediately Ewing organized another expedition, and evidence built up as to what really was happening to the sands and sediments and the abyssal plains. That was one of Ewing's greatest contributions, and is not often mentioned. His approach to the exploration of the ocean was really a completely new concept and has changed knowledge of the oceans more than any other thing. He didn't try to exhaust an interesting feature in one expedition — he looked at as many things as he could. Each expedition was a reconnaissance. Instead of shooting his wad on one voyage and spending perhaps the rest of his life working up the data, he went back and got more. He tested his theories at sea, and when he announced them they were right."

Ewing mounted two expeditions in 1948. *Atlantis* explored new stretches of the Mid-Atlantic Ridge, though Ewing flew home to be with his father, who had had a heart attack. In September, he brought *Atlantis* back to Woods Hole, carried a load of data back to Columbia, and met his classes. *Atlantis* sailed for the Cape Verde Islands, with David Ericson and Bruce C. Heezen as co-chief scientists, and crossed a new

85

Coring device ready to lower.

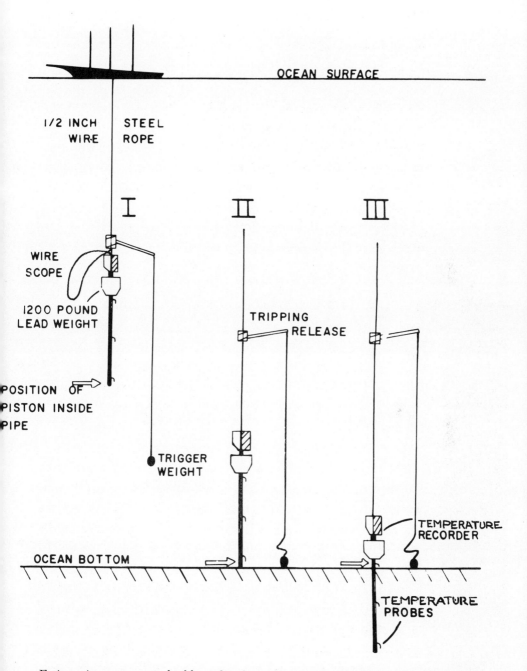

OCEAN SURFACE

1/2 INCH STEEL
WIRE ROPE

I   II   III

WIRE SCOPE

1200 POUND LEAD WEIGHT

POSITION OF PISTON INSIDE PIPE

TRIPPING RELEASE

TRIGGER WEIGHT

TEMPERATURE RECORDER

OCEAN BOTTOM

TEMPERATURE PROBES

Ewing piston corer works like a hand pump. As the tube falls, the plunger inside sucks in the mud.

abyssal plain east of the Mid-Atlantic Ridge. This voyage was an anthology of oceanographic distress. A magnetometer, which *Atlantis* was the first ship to tow across an ocean basin (the Ridge was intensely magnetic), broke down continually. *Atlantis* seemed tired. The electric brake of the coring winch threatened to burn out, and after a few cores the chief engineer refused to run the machine; the chief scientists radioed for a new part. The operator of a smaller winch hoisted the deep-sea camera at full speed out of the water and against a pulley; the wire parted and slashed his shoulder, and the camera went to the bottom. A trawl net mysteriously sailed off and vanished. "They took it in the whaleboat with the TNT, for a biologist aboard who wanted to catch fish," says David Ericson, "and they had hardly lowered it three hundred feet when the floats that supported it took off across the surface faster than they could sail after. It was mystifying at the time; someone suggested a big shark or a whale had tangled in the net. But clearly the trawl had simply hung down into a fast-moving countercurrent; if we'd realized such a thing existed in those days we'd have been heroes."

The mainsail ripped, and was resewn. When *Atlantis* got to the Cape Verdes, the winch part was not there but at Dakar on the mainland — from which a sailing packet was the only transport. Again belching greasy smoke from bunker oil, she went on to Dakar, then sailed west for the Caribbean. Southwest of the Cape Verdes she stopped for a core sample. Just before the corer reached bottom, the cable broke near the surface and the short end whipped across the deck. No one standing about was in its way, but two and a half miles of wire went down with the coring tube. As *Atlantis* came over the eastern foothills of the Mid-Atlantic Ridge, the mainsail ripped again; the old piece of canvas, scarcely fit for further use, was put away without repair. No other was aboard. *Atlantis* proceeded

When the core is back aboard, the mud is pushed out of the tube . . .

. . . except when there isn't any. The tube landed on a manganese nodule.

under power alone, and, next day when a seaman developed an acute stomach pain, at full speed. Barbados was three days away. During the night there was a loud crash, and the engine began to race. The propeller had broken loose on its shaft. *Atlantis* limped ahead under what sail she had as lights were rigged on deck and a bosun began stitching the tired mainsail. The sick crewman became delirious. During the morning the propeller fell off the shaft and sank. By noon the mainsail was set and *Atlantis* was making ten knots before a fine breeze. In the evening the sail ripped again; the sailmaker got out his needle. Later the mizzen ripped. *Atlantis* did reach Barbados, and the sick man was put in hospital apparently neither worse nor better. Ships' propellers were not in stock on the island, but the mainsail was reinforced there and *Atlantis* sailed for the graving docks of New London. On the way, the barometer began to drop alarmingly and reached a low the captain admitted he had never seen before; with all sails furled and double-lashed, *Atlantis* heeled over so far before the subsequent squall that it was easier to walk on bulkheads than decks. South of Long Island the mainsail ripped a last time, and she limped into New London under short sail three days before Christmas. On the new year Ewing moved into new quarters.

For several years, since he arrived at Columbia, Ewing had tried to share with his employers the perception that there was little future trying to outfit ships and operate seismographs in three small rooms beside Amsterdam Avenue. Early in 1948 he had an invitation from MIT to start a school there for geophysics and the study of the ocean floor; the offer included a large estate on the coast, a full professorship, and positions for his group of twenty-odd associates and students. "Ewing's reaction to the MIT offer was interesting," says Frank Press, who was one of the students. "A group of us came up with him to hear MIT's story. When we'd heard them out, he took us off into another room, and after a very short discussion he had each of us vote on a piece of paper whether we were to come or to stay at Columbia. I don't think many men would have done it, but it was perfectly natural for him, just a part of the way he regularly treated us. The vote was unanimously no." Later Ewing liked to say that MIT had never produced a Nobel prizewinner. Columbia offered him the directorship of Hudson Laboratories, a new institute that the university hoped would make of sonar and marine acoustics what MIT's Lincoln Laboratory had of radar. The offer — largely classified research in a narrow field Ewing had already left behind him — was not well calculated. The university also had the disposal of Thomas Lamont's country place, Torrey Cliff, a large house, outbuildings, and a hundred and twenty-five acres on the Hudson Palisades, that the banker's widow was offering with a few suggestions for its use. President Eisenhower and the medical school expressing no interest, Ewing got it; the university accepted Torrey Cliff and a half million dollars to get something going. Ewing agreed with the university that Lamont Geological Observatory — a subdivision of the Department of Geology, which it soon dwarfed — was to be self-supporting, whether by endowments or by contracts for research. He was cautiously urged to treat the half million he had as endowment and spend only the income. The principal was about a year's operating expenses, and Ewing cautiously spread it over two years. If in two years, he said, Lamont had not done work that would attract more money, the place should be closed down.

"My chiefest worry was about the fractured condition of the Palisades basaltic sill. When I knew it was a deal I rushed right out with a portable seismograph and set it up in Lamont's pool — which was dry." The Palisades transmitted earthquake waves without distortion; Ewing's chiefest headache was his new neighbors in the surrounding settlement of Sneden's Landing, a refreshingly rural spot only half an hour from Times Square, inhabited by well-heeled and sensitive commuters in media. Ewing was regarded by some as a noxious heavy industry. (Only a few years ago, Lamont erected and dedicated an Instrumentation Laboratory, after failing to get a town zoning variance for a machine shop.) He had scarcely set foot on the Lamont place — on the first of January, 1949 — when a local politician nailed a cease-and-desist order on the door. "I *thought* it threatened to put all the trustees of Columbia University in jail. I was sorely tempted. It seemed a dandy idea."

Ewing and his micropaleontologist colleague David Ericson moved right into the big house at Torrey Cliff, Ericson because he had no teaching commitments on Morningside Heights, Ewing because he could not wait. The mansion, pink sandstone with red clapboard wings, is up a steep winding drive from Sneden's Landing, set among fruit trees and azaleas and magnolias. Behind hedges there are flower beds and vegetable gardens. Thomas Lamont had had a staff of gardeners; Ewing's energetic colleague Worzel presently lined up help at a nearby missionary college. Ewing made himself a study on the second floor out of the Lamont bedroom, and later annexed the breakfast room. Ericson installed his sediment cores and microscopes in the dining room and admired the chandeliers and the birds on the splendidly painted walls. Everyone else stayed in town until summer. Then geochemists set themselves up in the kitchen with their radiocarbon dating still, a dense thicket of fine glass tubing that sprouted out of the sinks and rose into the stove hood. Geophysics and submarine geology went to the second floor, gravity and deep-sea cameras to the third. An oil furnace was installed and the coal bins converted to an electronics lab. Cores were stored in the seven-car garage, the indoor swimming pool was boarded over, and the tennis courts presently built upon. The greenhouses were made into the first machine shop — without permission. Seismograph styli were washed in Lamont's tub. For some years, now, Lamont's seismograph station has been the largest in the world; it includes installations all over the globe. On several floors in the six sprawling new buildings at Palisades (all larger than the original house), one is apt to run into rooms, brightly lit day and night, where banks of seismographs are humming just audibly, their electric pens scratching paths across glacially slow-moving drums of paper. They sound like wainscotfuls of small mice. In the spring of 1949, however, before even the original seismograph station was established in Thomas Lamont's root cellar (which, during the depression, he was said to have stocked with enough goodies to tide him through a revolution he anticipated), Ewing found on his desk one day a three-by-five card from Frank Press, who was required to report daily in that format on the seismic news of the world: "No seismogram today. Worzel took the seismograph."

Drawing boards were set up over tubs, and file cabinets over toilets. There were, with graduate students, twenty-four on staff, and each year added about twenty more, until the situation in the house became desperate indeed — and was solved by the equally desperate, but temporary, expedient of keeping twenty or so constantly at sea. When an administrator was hired he was found an office in the old downstairs ladies' room. A scientist who arrived in 1952 had to search for a place to sit, finally finding one in a cranny called the butler's ironing room; he borrowed a chair, appropriated some packing cases to shelve his books, and went to work. When it snowed, someone got out a jeep and the wooden plow that in Thomas Lamont's day was worked behind a team. There were no assistants. Heads of departments scratched for their own data, and new students had them for shipmates. In the close quarters of the house ideas were exchanged freely and often, generating more. "The parties — every Friday in the dining room after work — were kind of good then, too," says an early resident. "Kind of noisy and wild. Geophysicists are a rather rowdy lot, at least the ones you meet in the field. And cruises were a satisfactory supply of West Indian rum."

"At the beginning," says Frank Press, "each of us took on a major field. Marine geophysics was shared by all. We wanted, with Lamont, to get into different kinds of work than we had been doing before. We approached new fields with confidence; we read everything and knew what had been done. From the beginning there was the twenty-four-hour-a-day, every-day-of-the-week commitment to work which characterizes Lamont as against other institutions. We worked full time in our different fields and Ewing worked full time with each of us. I took on earthquake seismology. We developed new instruments and looked at new parts of the sound spectrum. We brought our ocean experience to bear on earthquake seismology. Ewing hadn't worked in the field before, but he was a well-trained physicist and as sophisticated an instrument designer as there was. He also wanted his students to be, and he would make something and tell you to go make another. We got old throw-away instruments wherever we could and made them into something else; and we had this from Ewing, that once we saw an instrument we knew how it could be improved. He was a very tough boss, and a mistake was unpleasant to make because he'd either make fun of it or do the thing himself. But once he had confidence in a person you got a lot of freedom and support. Our papers were written sentence by sentence together. He had tremendous drive. He has a competitive na-

Sediment cores in the dining room; geochemists in the kitchen.

ture and wants to be the first to discover anything. When we'd run into an obstacle his response would be, 'There must be a solution, there has to be a solution, we're just not being clever enough yet.' Some of us would have been ready to give up."

"He'd stomp around his office yelling, 'Why doesn't someone do something about this,'" says a colleague of those days. "He couldn't stand incompetence, and next to him practically everyone was incompetent. One day I went out for a long lunch with the key to the machine shop in my pocket. It was unforgivable. When I got back everyone was waiting for me. Doc never said a thing. He picked up a piece of scrap metal, went to the bandsaw, made a big jagged star, made a hole, polished it, found some small chain, and handed it to me."

"Our contributions," says Press, "were rapid and important, and in three or four years we were competing with thirty-year-

95

old laboratories. Our secret? A diverse approach, elegant mathematical solutions, hard work, and the equipment we built ourselves." In some circles, the haste with which Lamont began establishing fundamental truths about the earth and ocean floor was thought presumptuous.

"One day in January, a week or two after I moved into the Lamont house, Woods Hole called and said they had two ships going to Bermuda next month and no one using them on the way. Would we like the time? February is a bad month to operate in the North Atlantic, but there was no question what the answer would be. We had been trying for fifteen years, since I first went out on *Atlantis*, to discover if the continental basement extended out under the ocean. We had shot seismic profiles between a ship and a whaleboat in deep ocean, but our shots lacked the separation and the explosive power we needed to reach basement. Getting two ships to work together in the ocean has been one of the themes of my life. I didn't go on the expedition. It would have been truancy, and I was still a new boy at school. Joe Worzel and Frank Press, Brackett Hersey from Woods Hole, and Gordon Hamilton, who ran our Bermuda station for many years, went. They got two good seismic refraction profiles in two weeks at sea. The rest of the time it was too rough to work. When we had the data worked up, we got answers that we thought were accurate, the first time we had figures for the nature of the earth's surface in deep ocean that we could really swear to, though we'd been after them since before the war. We knew how much sediment there was and how much rock underneath, and what kinds. But out in the middle of the ocean our two little profiles looked kind of lonesome. Among the thoughts we had, Worzel and Press and I, was it would be awfully nice to get hold of more ships and measure beyond Bermuda, because maybe what we'd seen wasn't typical of the whole ocean — it certainly wasn't what anyone expected. We drew a line one night, Worzel and I, to Bermuda and two hundred miles beyond. And I said, 'If in a lifetime we can get enough ship time to make a line of stations to there, my life will be complete.' Every now and then Joe and I talk about it when we look at our ships' tracks on the world chart. I didn't dream we could ever look at any appreciable part of the world seismically at sea. We couldn't conceive that we'd ever get to the Pacific and Indian oceans. There wasn't that kind of money, or ships, in our lives then, or anywhere that we could see. We weren't much concerned in those days with theories. We were principally concerned with the question, What is the earth; and there weren't many answers. We were in no position yet to even wonder how it got that way.

"That's how Lamont, and I, got into earthquake seismology in such a big way — because we didn't think we'd ever have enough ships to do the job we wanted to do at sea. I wanted to look around, not spend my life looking at one station. We tried, Press and I, to think of the implications of our seismic profiles for earthquake wave propagation, and did our investigation of the world's oceans in a hurry. We found how to use a part of the seismogram called the coda. The name tells you how much seismologists thought of it. It consists of beautiful great big swooping waves — called surface waves — and was universally considered useless. When it appears, it's the largest part of the seismic record; and, because the coda only appeared when an ocean lay between the source and the recording station, it was thought to be an effect of the water. I saw my first coda during the war when I used to go down to Washington for meetings. The train back to Woods Hole was at midnight, and I would go around to the Coast and Geodetic Survey office before five and spend the evening reading seismograms as you might detective novels. They didn't have very good records, but they had lots of them and from many parts of the world. I got real curious about that coda and I saw a lot of it. Then I'd come out before midnight, trying, because I had to get past a guard without having a pass, to look as much like a tired, dyspeptic bureaucrat as I could. That wasn't always any trouble.

"It was natural enough to think we could do the same things to surface waves that we had done to explosion waves; but where we'd been working with waves of thirty cycles per second, earthquake waves might come several seconds per cycle. It was a matter of scaling things down seven or eight octaves. Unfortunately, a distinguished English mathematician, Stonely, had 'proved' years before with excellent mathematics that surface waves were not influenced by the oceans. Stonely, however, had had entirely mistaken and anachronistic notions about the depths of the oceans, and these had led him to false conclusions. We designed and built seismographs to get surface waves, and began to accumulate records. We had all the arithmetic from experiments we'd done during the war for the Navy on the propagation of sound waves in shallow water. All we did — in one night — was plug in the different units. And they fit. That was our paper. It was a poor man's way to cross lots of oceans.

"The surface waves told us that the figures we got on that little cruise to Bermuda — for the nature and thickness of the sediments, and the nature and thickness of the basement, right down to the top of the mantle — were typical of all the ocean basins. We were able to show that, contrary to the prior assumption of almost all geologists and geophysicists, the ocean basins are all alike; that there are basically two kinds of crust on the earth, continental and oceanic; and that there is a remarkable consistency within each class. The continents were known to be essentially granite, and quite thick; the mantle is about twenty-five miles down. Our profiles from the North Atlantic showed an ocean crust only three miles thick; and this was wholly unexpected. The wave transmission in the oceanic crust was consistent with basalt, which is denser and heavier than granite and exceptional on the continents. A few men had been enterprising enough to suggest there might be some basalt under the ocean. We found the oceans are all basaltic. Our profiles on the way to Bermuda also showed forty-five hundred feet of loose, unconsolidated sediment, and three miles of water. The surface waves showed, above the crustal rocks, an ocean deeper than three miles. The obvious answer was that the increment was the loose sediment. Our results were typical of a great number of wave paths throughout the Atlantic Ocean. We were rather puzzled to get almost the same figures for the Pacific — the Pacific averages a thousand feet deeper than the Atlantic. We had to assume there was less sediment in the Pacific, nearer half a mile than a mile; and, as it turns out, there is."

The proof was elegant, and deceptively simple, with pleasing symmetry, and about as basic and elementary a matter as could be. "The completeness of the segregation of the sialic (granitic) rocks into continents is underscored as one of the major problems of geology," Ewing wrote in the fifties. Scarcely anyone was listening then, but today the problem is much discussed. Equally, the existence of the paper-thin ocean crust is one of the most important things known about the earth; geologists have speculated that it is the chilled surface of the earth's mantle. Today geologists see it as basic to the earth's creation and renewal of its surface in earthquakes, eruptions, and mountain ranges. But again Ewing had the experience of discovering a significant fact and finding no one prepared to make use of it. The continental ghetto, the singu-

lar ocean crust and its extraordinary thinness, might seem
suggestive enough to have raised a storm of ideas and of the
theorizing and speculation that are not uncommon in geology.
Perhaps, however, they were too powerful stimulants, too suggestive, and the mind, instead of responding, boggled. "We
were kind of naïve then," said Ewing. "We thought you discovered something and published it, and the world would
know about it; but several years later even leading seismologists hadn't heard of our paper." The brutal fact was grappled
with by a couple of marine geologists after more than a decade, and by geologists after two. In the meantime, a lot of
geology needed to be rewritten. Many geologists had claimed
there were ancient continents sunk beneath the seas; a leading
authority had said the ocean floor was "nowhere more than
fifty miles thick"; and distinguished seismologists had described all the oceans but the Pacific as made of granite like
the continents. None had the benefit of good data. A variety of
theories, now untenable, needed to be scrapped, a treatment
that was not always administered. Ewing said: "There was the
interesting doctrine about that the Pacific is different from the
other oceans, the only 'true' ocean, maybe made by the moon
flying out of the earth at some early date; and that the Atlantic
and others were just sort of low-lying continents. Well, they're
just like the Pacific. So if there was an Atlantis it's on the
moon." The moon out of the Pacific was a story started by Sir
George Darwin, son of Charles, and, not long after Ewing and
Frank Press announced their findings about the oceans, it was
hair-raisingly retold in one of the most popular books written
about the ocean: "This rolling, viscid tide [gathered] speed and
momentum and [rose] to unimaginable heights." Unimaginable
indeed. All the granite of the Pacific Ocean basin supposedly
flew off to form the moon, which is not as dense as the earth.
The testing of this moonshine, according to the *New York Times*

a few years ago, would be one of the scientific rewards of putting men on the moon.

In Ewing's office is a letter, signed by the leading Japanese seismologists of the day, complimenting him on the solution of surface waves. "It may turn out to be the most important work I did," he said. In the next decade, he and Press and others built surface waves into a new science and the most powerful means of learning about the earth's crust and upper mantle — where earthquakes, volcanoes, and mountains, for whatever reasons, have their origins. Ewing and Press showed that earth waves, water waves, or atmospheric waves can create the other two, something they called coupling. They showed that a water wave recorded in England after the Krakatoa explosion was not a spent tidal wave but the product of an atmospheric vibration; and in 1954 they found wave coupling in front-page news. On a clear, calm, sunny day on Lake Michigan two fishermen were washed off a pier by a single freak wave and drowned. Though a tidal wave was blamed immediately, there had been no earthquake; and Ewing, Press and William Donn showed how a weather front crossing the far end of the lake had raised a small wave which resonated and grew as it traveled toward the pier. Lamont recorded waves from both atmospheric and, later, underground nuclear tests. Surface waves can be used to distinguish explosions from natural events, while coupling has been used in the hope of disguising explosions — atom bombs have been set off in large caverns where the surrounding air absorbs some of the shock, decoupling it, as the phrase goes, from the ground (the ruse fooled no one as much as its own advocates). Lamont's hydrophones in the Atlantic off Bermuda have recorded water waves made by ground waves made by air waves made by Apollo rockets. During the fifties Ewing established seismographs in remote parts of the world to get better records, especially of surface waves. In 1956 there was an earthquake in Orleansville, Algeria. "We made money in both directions on that one," he said later, still pleased. "In the Mediterranean there was a turbidity current, something we'd been waiting to see, and our station in Cape Town got one of the longest purely continental wave paths that had been recorded." To learn about the earth's mantle, the layer under the crust, Ewing and Press began building machines to record longer surface waves. (The longer the waves, the deeper into the earth they reach.) By 1959 they had picked up waves with a frequency of one per thousand seconds; each wave was about the length of the United States. "We learned in those days how Rayleigh waves worked; computers were just coming in and as fast as they improved them we had bigger problems. We would conceive of a paper and write it in the same day. We lived through a period when earth sciences changed from an immature, descriptive field to a mature body of knowledge. The free vibrations of the earth were a wonderful story. Free vibrations are the longest vibrations an object can make, like an unstopped cello string, or a bell. People had wondered if you could make the earth ring like a bell. For years imaginative men would look at little wiggles and say, 'Maybe that's a free vibration.' We kept making longer and longer period seismographs, and getting longer waves. Finally in 1962 there was an earthquake in Chile. The earth did vibrate like a bell for over twenty-four hours after, with waves which were four thousand miles long. We, and four other labs, all reported them in the same issue of the *Journal of Geophysical Research*. It's exciting when a science is that active."

Six

That Man Is Taking Too Many Cores

ON HIS WAY HOME from the Mid-Atlantic Ridge in 1947, Ewing had radioed to Henry Stetson in Woods Hole — to whom he intended giving the sediment cores that he was taking — and asked where he could most usefully core between Bermuda and Cape Cod. "I got back a message, 'If at the end of a two-month voyage you attempt to take any cores at all, you will have mutiny.' I didn't think much about mutiny. But later, when I tried to give our cores to Stetson, he said, 'No thanks, I've got too many.' So I thought I'd better study them myself, or eat them. He was only studying the top centimeter anyway." During the winter Ewing got excited with the cores, and for the two cruises of 1948 he had a new coring device of his own devising, essentially today's model. He wedded the best features of existing devices, the simplicity of Stetson's and (to overcome the friction of the tube and draw the mud farther in without its being squashed) the piston of a complex Swedish gadget he'd read about. His own contribution was characteristic. Both Stetson's and the Swedish tube were turned on lathes; they were so expensive that their creators were reluctant to have them used, and perhaps wrecked, anywhere that might be hard, like hills, mountains and steep slopes — most of the interesting places. Ewing made his corer with two-and-a-half-inch steel boiler tube, and did not skimp the ship for spares. He set out to take as many cores as he could from as many kinds of topography as he could find on the echo sounder. There were oceanographers who did not think what he was doing was a good idea. "In those days," an old Woods Hole hand says, "when someone took a core, that represented a year of working up the data. If you took a dozen cores, you were tied up for a long time. Ewing took up coring with a vengeance; and by the time he'd got a couple of dozen, a lot of scientists were pricking up their eyebrows and saying, 'That man is taking too many cores. He's not working up his data. Why is he being given money to take more?' But what Ewing had noticed right off was some of the gross features of the cores — I remember he said, 'When you can show me two alike I'll stop and study them in detail' — such great differences that there must be some outside influences at work disrupting the normal course of sedimentation. It was these gross differences, the occasional gaps in the sediment column, the occasional layers of sand, and others, that he set out to explain by getting enough cores to see if a pattern could be found." Ewing took twenty-seven cores in 1948 before the second expedition lost the coring apparatus. At one time in 1950 he had two ships out at once and the chief scientists were trying to see who could take the most cores. Many of these cores were full of sand; and sand was not, as Ericson and Ewing later wrote: "in accordance with orthodox theories of sediment in the deep ocean basins." It was, and still is, an axiom of geology that sand cannot form in the ocean, much less in mid-ocean. It is ground up in streams and carried by them to the sea, where tides and waves and some currents spread it along the shore. Sand — beach sand, coarse, clean, chiefly bits of quartz — had been found in mid-ocean before *Atlantis*-150. The *Challenger* expedition found it once, wondered why it was there, and concluded it wasn't — that the apparatus was contaminated before lowering. Later expeditions, however, found more; sand became a perennial oceanographic puzzle, impossible to explain or explain away, though oceanographers and geologists were wonderfully inventive. Something called sedimentary vulcanism was supposed to have thrust sand up through the ordinary abyssal sediments from below, where it had lain since being the beach of an infant ocean on an infant globe; drifting icebergs to have transported it from the shores of Greenland; and Pleistocene windstorms to have blown it out from the Sahara. The roots of drifting trees and droppings of migratory birds both were invoked in explanation. In 1920, a scholarly German reviewed all the theories in one chapter of a book called *Geologie des Meeresbodens*. He elected as most plausible one advanced to explain sands cored by the German 1901–1903 South Polar Expedition: that the sand had been deposited in shallow water and that the land thereafter sank to abyssal depth. Alas, still in the twenties, a core was taken in which there was abyssal sediment below, as well as above, the sand. It became necessary for the ocean floor to have heaved itself up several thousand fathoms to make a beach, and then — perhaps wilting under the strain — ponderously to have sunk back again. In the nineteen forties, the accepted doctrine was a catastrophic drop in sea level, in the course of which the sands were deposited on beaches and several other mysteries obligingly cleared up. Ewing guessed, rather more modestly, during *Atlantis*-150, that the sand might come off the tops of neighboring seamounts — seamounts were a relatively recent discovery of wartime. A far smaller, and more feasible, drop in sea level would turn these into islands, and near one sand core *Atlantis*-150 found a seamout.

"One day at the end of May, 1949 — just a few months after we moved into Lamont and while we were just starting to collect surface waves on our seismographs — Woods Hole called and asked would four days at sea starting the day after tomorrow in Jacksonville be of any use to you? I wanted to core the Blake Outer Ridge off the shore there, where they've just discovered they'd like to drill for oil, and we flew our coring apparatus to Florida. It nearly broke my heart to ship a ton of lead by commercial plane. We'd only been half a day on station when quite a storm came up and the captain hove to to ride it out. When it passed, he came to me with a kind of smirk and said that as it happened he hove to in the middle of the Gulf Stream and we were now only a few hundred miles south of Cape Cod; and that since we hadn't time to go back on station we might as well go straight on in to Woods Hole. My immediate answer was, 'Not on your goddam life.'

"We decided instead on tracing the Hudson Submarine Canyon as far out to sea as we could. There had been a long controversy between Reginald Daly, of Harvard, who had proposed that submarine canyons were made by avalanches of muddy water, and Francis Shepard, of Scripps, who said they were made by rivers in an ice age when sea level had been lowered enormously by glaciers. The year before, we had crossed a wide notch in the bottom about halfway to Bermuda, which looked to me like the cross-section of a submarine canyon. If we could follow the Hudson Canyon that far, into water three miles deep, we would have a fact of some magnitude, which any theory would have to account for." Generations of fishermen and navigators knew the Hudson Canyon as a ditch fifty to a hundred feet deep and a mile wide, a useful seamark, crossing the continental shelf from the river's mouth at Sandy Hook; it runs southeasterly for about a hundred miles to the edge of the continental shelf, where the water is about a hundred fathoms — or six hundred feet — deep. Similar canyons on other coasts about the world also had been known for years, always at the mouth of a major river — the Congo, the Ganges, the Amazon. But in the mid-thirties, during some of the first echo sounder surveys, A. C. Veatch, a consulting geologist, and Paul A. Smith, of the Coast and Geodetic Survey, found that the Hudson Canyon turns into a colossal gorge at the edge of the continental shelf. They followed it for the very short distance it took to fall from a hundred fathoms to a thousand (of which the canyon itself accounted for over half), the limit of their survey. The early echo sounder surveys vastly multiplied the number of canyons known. Along the Atlantic coast of North America from Maine to New Jersey, every fifteen or twenty miles, there is a submarine canyon fifteen hundred to two thousand feet deep and five to ten miles wide, running out to at least a mile and a half of water. The origin of submarine canyons became a major puzzle. For quite a few years oceanographers were hypothetically drying out the oceans to account for the canyons (and deep-sea sands); it seemed a reasonable theory, though it was wrong, and the oceanographers seemed quite at ease disposing of substantial quantities — even a mile and a half — of water. Reginald Daly, who was a progressive sort of thinker, realized that lowering sea level thousands rather than hundreds of feet was a formidable job of engineering — indeed, a catastrophe far in excess of what it was required to explain. Some more modest catastrophe was required. Daly proposed what he called a "density current," something that had been imagined as long before as 1885, by a Swiss engineer named Forel who discovered a canyon in Lake Leman. The Swiss was ignored, but Daly was sufficiently eminent for the idea to get some attention — almost entirely devoted to dismissing it. Daly imagined that, as the ice ages lowered sea level a modest amount, the soft sediments of the continental shelf were exposed to surf and got churned up in the water; and this abrasive mixture ran downslope, wearing away its path. Other scientists doubted that. Such a current had been observed once in nature, in Lake Mead behind Boulder Dam, but was dismissed as the product of exceptional conditions. In Holland, after the war, Philip Kuenen produced turbidity currents — as he, and others since, called them — in a laboratory tank, and made impressive photographs of the flows in progress. The results he described included a flat surface where the turbidity current deposited its sediment, and layers of sediment graded from coarse at the bottom to fine on top; but Kuenen could not produce substantial erosion. Geologists and oceanographers like Francis Shepard (who was one of the most eminent oceanographers of the forties and fifties, and who had made a specialty of submarine canyons) therefore continued to believe that sea level had been lowered a mile or more in relatively recent times. They felt quite strongly about it; one scholar spoke with fine irony — in his presidential address to a learned society — of the "galaxy of mental stars" who thought they had better ideas than lowered sea level.

Atlantis crossed the Hudson Canyon near the edge of the continental shelf and zigzagged out to sea over it. Before reaching even a suggestion of an end, she had to return to Woods Hole. She was then a hundred and five miles beyond the edge of the continental shelf, at the foot of which the

Canyon had been supposed to end. Two miles down in the abyss, the ocean floor had flattened out after the abrupt drop of the continental slope, but the Canyon remained larger than any portion of the river supposed to have cut it. When *Atlantis* turned back, the Canyon still was three miles wide and nine hundred feet deep, and was in more than eleven thousand feet of water. This was a fact of some magnitude, and another short cruise on *Atlantis* was promised for the fall to finish the job. All the time of several ships was contracted for until then, but in July Ewing and a Lamont group on *Atlantis* discovered a new and unique submarine canyon. It was in mid-ocean. Though everywhere under more than two thousand fathoms of water, its sinuous course and leveed banks were reminiscent of an old river bed. Several hundred feet deep, several miles wide, and (as more surveys showed in succeeding years) at least a thousand miles long, the new canyon was named, with childlike directness, Mid-Ocean Canyon. It runs along the ocean floor south from the Cabot Strait between Greenland and Labrador into the eastern edge of the great abyssal plain discovered in 1947, when *Atlantis* sailed for three and a half days over the part between Bermuda and the Mid-Atlantic Ridge. In late summer, one of Ewing's colleagues sent to press a paper noting several riverlike features of the Hudson Canyon and describing the abyssal plain as the lowest of several types of former sea-level terraces in the ocean, presumably dating back to the birth of the ocean itself in the early days of the world.

"After we found the flat plains," said Ewing, "all winter the main topic of conversation between me and the graduate students, whenever anyone was going to the airport, whenever there was time, was how could anything get that flat. Our speculations went pretty far; we didn't try to make them reasonable, just to consider anything that could have made the

degree of flatness that we had observed. There wasn't an echo
sounder worthy of the name until we built one, for the first
cruise of *Vema*, in 1953; but we did have seismic reflection
measurements every half hour that were accurate within a few
feet. We knew the abyssal plain was as flat as any alluvial
plain. We even said maybe the whole ocean dried up. But one
thought emerged — maybe old Kuenen's process was doing it.
He had not thought that turbidity currents were oceanwide,
just small things in lakes and ponds — but he had got a flat
surface. After that spring cruise down the Canyon, I was wild
to get back there and show what was in it. In August I finished
early with some seismic work for the Navy and took *Caryn*
back to the Canyon. She was a poor little ship without an
adequate echo sounder or winch, but I got cores from the floor
of the Canyon, the walls, and the adjacent ocean bottom that
proved definitive. In September, with the echo sounder on *Atlantis*, we followed the Canyon out across the Gulf Stream. It's
very difficult to show what is there, because your navigation
goes to pot in the current, but we did." The Canyon got gradually smaller farther out to sea until it was lost in a broad plain
sloping gently to the southeast. The plain was the same one
from which *Atlantis* had gotten some cores containing layers
of sand in 1947 and 1948; it proved, in fact, to be the western
end of the abyssal plain that stretches north of Bermuda and
southeast of Newfoundland to the foothills of the Mid-Alantic
Ridge a thousand miles away. Just beyond the end of the Canyon was a small seamount. On the echo sounder record its
flanks emerge abruptly out of the flat sea floor, without transition, making a crisp angle — as though it were partly buried in
the plain. It was named Caryn Seamount. The Hudson Canyon, when it was lost in the plain, was 200 miles beyond the
foot of the continental slope, 350 miles from Sandy Hook, and
2,650 fathoms — 15,900 feet — beneath the surface. "That set-

tled it as far as I was concerned," Ewing said. "But I was kind of naïve in those days. A lot of men weren't interested in changing their minds. As late as the International Oceanographic Congress in 1959, Francis Shepard read a paper to show there was no such thing as turbidity currents. I had to restrain myself from asking to have his slides shown over again as rebuttal; he had perfect evidence to support our conclusions."

The cores Ewing took in August were analyzed ashore by David Ericson, the micropaleontologist. Cores are given a superficial description as soon as they are brought back — length of core and of any visible layers, color, and type of sediment. Later the core may be sampled every ten centimeters and the microfossils examined — and sometimes counted — for a chronology of the core. Cores from the walls of the Canyon were of Eocene sediment forty million years old. Cores from the Canyon floor had an uppermost layer of about fifteen thousand years' worth of Recent deep-sea ooze — representing the interval since the last ice age — in which were a few pebbles from the Canyon walls. Beneath the Recent abyssal ooze were layers of sand and shallow-water fossils, all characteristic of the continental shelf southeast of New York. The last core was the most expressive. It was taken adjacent to the Canyon, from the surface into which the Canyon had cut, and was composed entirely of abyssal ooze and fossils deposited in deep water continuously over the last hundred thousand years, during which the Canyon was supposed to have formed. It and later cores just like it were unimpeachable witnesses that sea level had not changed greatly, much less left the area high and dry for rivers to flow. Yet the Canyon had been cut through those sediments.

In the summer of 1950, Ewing was on *Atlantis* doing seismic work, and when a second ship failed to show up, he took her

An experimental turbidity current of fine sand and water rushing down a model valley in a laboratory tank.

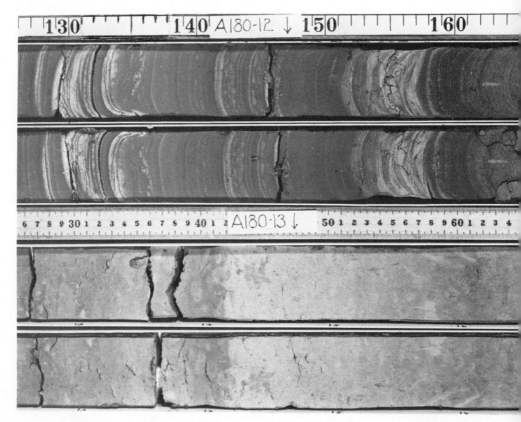

Core of fine sand and silt layers (some graded) from turbidity currents (*upper*); and core of normal abyssal ooze (*lower*).

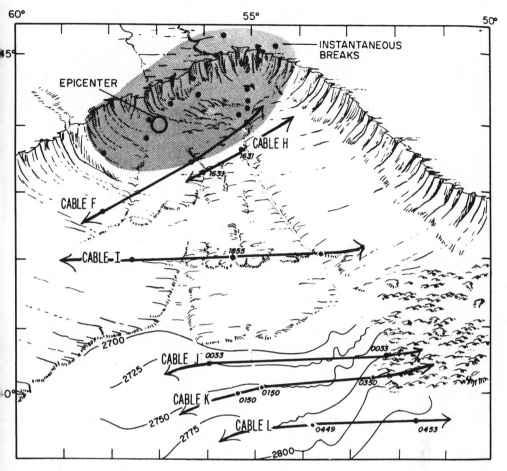

Continental rise south of Newfoundland, location of the Grand Banks turbidity current, showing times and places it broke telegraph cables.

Seismic profile of the landslide that started the turbidity current, with contorted sediments at its foot.

back near the mouth of the Hudson Canyon and took cores for several days. He got ten in an eighty-six-hundred-square-mile area of the plain that is over fifteen thousand feet deep. The long cores that his new device took included quantities of sand that were almost insulting. Nine out of ten contained sand, in layers that ranged from thin films to six meters thick. Some contained many layers. In one, the sand was on top — a hard circumstance to explain if the sand were also supposed to be very old. The sand beds have an extraordinary quality. They are sorted. The particles are perfectly graded, from coarse gravels at the bottom of a bed to the finest gray clay on top. In an article, David Ericson, Maurice Ewing and Bruce Heezen, who was then a graduate student, disposed of some of the old ideas about sedimentation in the ocean: "While we are quite willing to admit the Pleistocene storms were powerful, we think it is asking a little too much of even a Pleistocene gale to transport the [half-inch] pebbles of the Hudson Canyon gravel a distance of more than 100 miles. . . . Any explanation [which involves] greatly lowered sea level leads to the embarrassment of having to reduce the North Atlantic to hardly more than a series of isolated puddles." And the water would have to be pumped in and out of the ocean very quickly. A core taken off Bermuda showed "no less than seven alternations of graded shallow-water material with typical deep-sea ooze." Cores from the Caryn Seamount, and every other piece of high ground in an abyssal plain, show none of the turbidity current characteristics — layers of sand, graded sands, shallow-water fossils — but instead an uninterrupted accumulation of deep-sea ooze. In 1951, in over four thousand fathoms in the Puerto Rico Trench, the deepest part of the Atlantic Ocean, Ewing spliced *Caryn*'s winch cable to *Atlantis*'s and took two cores containing several graded beds of sand from the narrow abyssal plain at the floor of the trench; in the sand were remains of *Halimeda,* a tiny plant that lives in the surface waters along coastlines.

After months of gathering evidence of the passage of turbidity currents, Ewing and his colleagues had developed a strong urge to get the measure of one; and they concocted an ingenious "full-scale experiment in erosion, transportation, and deposition of marine sediments by a turbidity current" — as the reporting paper described it. "The Grand Banks earthquake of 1929 was the main topic of discussion in the first national meeting I ever went to, in the spring of 1930," Ewing said. "Some very funny things happened after it. I remembered the discussion, and I told Bruce Heezen, 'That must have been one!'" The Grand Banks earthquake had indeed had its curious aspects. The earthquake center was on the continental slope south of Newfoundland. Over a dozen of the transatlantic telegraph cables which litter that area had broken — not surprising when the quake also rattled windows in New York — but not all cables broke at the same time. Cables on the continental slope broke instantaneously. Cables on the continental shelf above the earthquake center did not break at all. Cables farther out to sea broke as much as twelve hours later and in two places at once. When repair ships went out, an expensive loss became even bigger, for the segments between the breaks could not be found, having apparently been swept away or buried. Lost sections as much as two hundred miles long were replaced. Beyond the continental slope, however, the cables lay on nearly flat ground — indeed, on an abyssal plain — and in several thousand fathoms of water. The geologist Walter Bucher had suggested that a tsunami broke the cables, and the seismologist Beno Gutenberg a landslide. It is generally agreed now that tidal waves are caused by submarine landslides, caused in turn sometimes by earthquakes;

but tidal waves gather their force in shallow waters, and landslides don't cross hundreds of miles of flat ground. Ewing and his student Heezen thought the story sounded a lot better with a turbidity current as the culprit. The earthquake started a landslide on the continental slope, where the cables broke instantaneously, but the landslide started a turbidity current. Water and mud roiled up together as the slide picked up speed, as snow and air do when an avalanche starts. Whereas mud and snow slides are restrained by friction and are slow-moving, turbid mixtures of mud and water, or snow and air, bound by no such restraints, will thunder downslope at enormous speeds; avalanches have been timed at over a hundred miles an hour. The Grand Banks earthquake dislodged a section of the continental slope — one can see the scar in records from seismic reflection profiling — a hundred and fifty miles long, eighty miles wide, and twelve hundred feet thick, stirring up a turbidity current of noble proportions. The first cable beyond the foot of the slope, fifty nautical miles away, broke fifty-nine minutes later. Six more cables broke one after another. Ewing and Heezen calculated that the current must have been traveling at over fifty knots when it broke its first cable and still have been going twelve knots when it went over the last, half a day later. Kuenen estimated the turbidity current was some two hundred and seventy meters high and five and a half kilometers long; its flow was six hundred times that of the lower Mississippi River. After twelve hours, four hundred miles from its origin, it passed the last cable (which was new, with a breaking strength of ten and a half tons) with sufficient force to carry it away completely over a two-hundred-mile front. It was then well into the abyssal plain. Three years passed before Lamont could get core samples of the area. For two years there were hurricane warnings, and *Atlantis* had to run to safety. The third summer, on *Vema*, Charles Drake, a graduate student, got cores along the route of the vanished portion of the last cable. They contained, on top, a bed of graded gray silt with shallow-water remnants, and, below a sharp boundary, normal abyssal clays. Four hundred miles from the earthquake center the turbidity current layer was a yard deep. Kuenen estimated the current had dropped a hundred cubic kilometers of sediment.

During the next several years, nature obligingly provided turbidity currents in different parts of the world, and Lamont Observatory documented them more minutely than the cable engineers unwittingly had the Grand Banks one. Turbidity currents became an established fact in oceanic science. The evidence Ewing and his associates brought for turbidity currents was accepted after a few years — it had to be — with certain holdouts. Francis Shepard wrote, "One looks in vain for evidence that turbidity currents actually erode." Börje Kullenberg, a member of the Swedish expedition that called abyssal plains lava flows, found on some outdated charts a topographic obstruction south of the Grand Banks that he said no turbidity current could pass. A professor claimed that turbidity currents were a far greater strain on the credulity than believing that half the ocean floor had sagged down twenty thousand feet and then heaved itself up again. "According to one man, no one could help but believe in lowered sea level while subscribing to sweet reason," says David Ericson with a mild air of astonishment. Other geologists, however, recognized that the relatively recent sediments Lamont was bringing out of the ocean were in many respects identical to much older stratified rocks they were studying on land. "Oil deposits in California show all the characteristics we were finding in cores of turbidity currents," says Ericson. In order for oil to form, a great deal of organic matter — chiefly the remains of little marine creatures like the forams — has to be very suddenly buried so that it decays in the absence of oxygen. It was difficult, before turbidity currents were observed, to produce such a burial even hypothetically. But turbidity currents clearly are responsible for the vast deposits of oil below the sea floor all over the world, which have so excited the oil industry in recent years. Early in 1974, Ericson and his associate Goesta Wollin succeeded in extracting petroleum from cores of turbidity current sediment, a finding of great significance for oil exploration (though it may be some time before the corner filling station sells boxes of mud). Certain ancient sediments had puzzled geologists for a long time. Along the Hudson valley near Poughkeepsie there are dramatic exposures of thick deposits of a hard shale, or dirty sandstone, called graywacke; the stone is in layers inches to yards thick, and the particles in it are sorted, from coarse sand at the bottom to fine silt on top. Geologists had been unable to find a mechanism to produce such graded beds until Kuenen's experiments. Kuenen went on to find turbidity current deposits in formations as much as three billion years old. Mountains are composed of beds that may reach total thicknesses of eight or ten miles, or even more. Almost all the layers are sedimentary, and most of the sediments marine sediments. It had always been hard to explain why sediments all the way down, even those once buried miles deep, are seen — when upheaval, folding and erosion expose them to view — to contain shallow-water fossils, as do the abyssal plains at the bottoms of trenches five and six miles below the surface of the ocean. During the fifties, Lamont expeditions found creatures at the bottoms of trenches which live only on shallow-water foods, presumably delivered to their door by turbidity currents. The odd layers and apparent climate changes that once baffled students of marine sediments are the souvenirs of turbidity currents. Turbidity current beds in any case are too thick for coring tubes to penetrate far into the past, and the search for old sediments has withdrawn from their reach, first to the hills, where the "endless snowfall" continues apace and ordinary meandering currents erode and redeposit abyssal sediment, then to slopes where the oldest stuff might have been exposed by landslide. David Ericson and others began piecing together a record of the world in the last several million years. The Pleistocene, the ice age, is of interest because man evolved during it, but in the fifties no one was agreed even on how long it had lasted. "Studying an ice age on land is like finding out how many times a blackboard has been erased," Ewing has remarked. "In the ocean it's like how many times a wall has been painted." By 1962, Ericson had a chronology of the Pleistocene glacials and interglacials. His Pleistocene began a million and a half years ago, longer than had ever been claimed before — some estimates were less than half a million years — and later studies have roughly confirmed or even added to his figure, allowing mankind a less feverish evolution than geologists had before. What turbidity currents have done to change much of the face of the ocean floor today, they must also have done on the ancient ocean floors that are now in mountains. Their sediment fills the deepest depressions and constantly spills over into larger areas until the pools of it join to make an abyssal plain. The abyssal plains have spread over huge areas of the ocean crust, burying and transforming the original topography. In recent years seismic reflection profiling has revealed the abyssal plains of seventy to ninety million years ago buried half a mile beneath the plains of today and beneath hills and dunes of sediment. They hint of conditions even more extreme, and more widespread, than those of recent ice ages.

Seven

I Wasn't Lonely a Bit

GOVERNMENT SUPPORT of science increased during the fifties; the National Science Foundation — which now dispenses most of that support — was established, and other agencies, especially the Office of Naval Research, became richer. In the summer of 1950 Ewing shot a line of seismic refraction profiles with two ships three hundred miles southeast of Bermuda, a little over a year after he felt his life would be complete if he could get two hundred miles beyond. (The oceanic crust was the same as west of Bermuda, and as in the earthquake surface-wave studies.) Just two or three years after its founding Lamont was competing with older institutions for a limited number of research ships. As Lamont needed more time for more projects, less was available. There was *Atlantis*, but

Woods Hole wanted more of her too. Ewing became unhappy. "We had a year or two of using random ships belonging to other institutions, and having them withdrawn from us on very short notice. The Navy recognized our needs, and we proposed to convert one of their ships to use fifty-fifty with Hudson Labs — across the river — who were working on underwater sound. The Navy told us to monitor the conversion; and we spent the better part of a year in Brooklyn Navy Yard and never had a day on her. They said Hudson Labs better have the first six months, and as it came our time, a meeting was held in Washington and some lieutenant commander got up and said the ship was being withdrawn for work of higher priority. In lieu of the six months, they gave us the money to charter — for three months — the *Kevin Moran*, an ocean-going tug on which we jury-rigged a heavy-duty winch. We took her to Newfoundland and West Africa, surveying the Mid-Atlantic Ridge and the new mid-ocean canyon, then to Brazil with *Atlantis*, on which we returned home. Joe Worzel had the luck and skill to find *Vema* and see her potential; and next year we chartered her, with an option to buy.

"I've often thought that the spirit in which we took that option can only be compared to saying, as you go back to New York across the bridge, 'Here's fifty-one cents for passage and an option to buy in ninety days.' The price was a hundred thousand dollars. I had no idea how we'd get it, but the option didn't cost much. Ultimately we bought her for the trustees without telling them. They were surprised, to put it mildly, that anyone could do it to 'em. One called me up the very same day and said, 'If this costs Columbia five cents it will cost you your job.' He found out because he was selling insurance to the university. I told him, 'If in five years you're not prouder of *Vema* than anything Columbia's got, you can have my job.' Joe Campbell, the university treasurer, believed in us, and the purchase actually amounted to his personal funds. When he took the option he told me: do research — he'd raise money. Two or three days before it ran out he called me and said he'd failed. I went back to my boys, and it was the one time they didn't back me up; they said that going on other people's ships for three months they spent all their time moving equipment on and off and didn't have any time to work — and that I had to get the ship. The last day of the option came and I called Joe at his office. He wasn't there. Well, where was he? They weren't supposed to tell. 'Tell me,' I said, 'or I'll be on the scene.' Then I telephoned Cooperstown where he'd gone to rest, and got his wife. He was playing golf and would be back at supper. That was too late. She said, 'If it's terribly, terribly important I'll get him.' I said, 'You'd better go,' and she went. There were two really heroic roles played that day. He heard me out and said, 'I'll drive down after authorizing the check.' *Vema* saved us time and saved us money; but we had to make a very big step to use her efficiently. We were geared to using someone else's ship three months of the year and studying the data the other nine. At the same time there were already five and six people in a room in that old Lamont house, and no place to sit. So we just worked everyone four times as hard for a while and kept some of them and the ship at sea for nine months of the year."

In the succeeding years Ewing, at least, became prouder of *Vema* than anything else Columbia had; and he learned the economic principle of keeping the ship at sea all the time, with minimum turnaround time, even before it became of prime importance in shipping circles. She was a three-masted schooner, two hundred and two feet long, slim and handsome. Her topmasts were gone, but she still had all three lowers. She had a fine aristocratic bow and an elegant stern; a few pictures show her under full sail, heeled over, rails buried in water. Her

Vema under sail when she was *Hussar*.

bowsprit, with a fierce eagle under it, went out thirty feet beyond the deck one way, the main boom about as much the other way. The tops of her masts were a hundred and twenty feet above the deck. They hung an immodest amount of sail, and *Vema* still holds the speed record for the passage from Montauk to Bishops Rock, Ireland, under sail. She belonged to a Nova Scotian, Louis Kenedy, who had found her in the mud behind Staten Island after the war, with watchmen selling the lead out of her keel. He hauled her off the flats and took her back to Lunenberg to replace his old ship *City of New York*. He couldn't resist a fine ship. But while *City of New York* had been built for Antarctic voyaging, too heavily for a profitable cargo ship, *Vema* was too fine-lined to make a good lumber schooner either. Kenedy put her up for charter, and Worzel found her through a yacht broker. Kenedy came with her as captain the first year, with a crew of his Nova Scotians. She had been so neglected before he got her that the wooden decks and deckhouses leaked on the geophysicists' bunks; some of them took tents to sea. But she was sound and could go anywhere with safety, if not comfort. She was built in the twenties of Swedish wrought iron — which is more enduring, and expensive, than steel — for Mrs. Marjorie Merriweather Post Hutton, and her husband, and sold when they were divorced. Ewing says: "Mrs. Hutton later became Mrs. Davies, and they built a ship called *Sea Cloud* which was three hundred and fifty feet long and had four masts, three of them square-rigged; when Davies became ambassador to Russia they loaded her with goodies and took her to Leningrad. We thought once of buying her for a second ship. *Vema* was bought by George Vetleson, a shipping man, who gave her her present name — she had been *Hussar* — and he sailed her straight back to Norway and I'm told took the king for a sail." Coming back to New York with only the crew aboard, *Vema* nearly sank. A

Part of the owner's quarters when *Vema* was a yacht.

hurricane took a queer jog and passed right over her as she tried to evade it. The barometer went so low the captain, a veteran of square-riggers, told the mate reporting that he must be mistaken; then it went an inch lower still (to 26.35 inches). Winds were reported at force twelve, the top of the scale. At such times odd things capture one's attention. The captain noticed the mate's ears were being blown forward and flattened against his cheeks. With bare masts, the ship lay over before the wind at forty-five degrees for hours, rail in the water. Word came from below that she was sinking. Her designers had provided ventilators hidden behind paneling on deck and in the cabins. The effect was pretty, but unseaworthy; they were pouring two-inch streams of water into the cabins and could not be covered. Another hour or two, said the captain, would have done for her. She had oriental carpets in those days, Louis XV bedrooms, and gold faucets; the antiques were the most of her damage. In 1941 the Navy took over all large yachts, in the national interest. Many like *Vema* were sent out on submarine patrol. "The idea," said Ewing, "was to creep up on the enemy submarine, without the noise of engines, and radio for a killer. But they gave the ships no radio or sonar gear." *Vema* was taken later to Kings Point and gutted for a floating dormitory; from there she went to Staten Island until Louis Kenedy saw her. With *Vema*, Ewing felt his laboratory was truly founded. "The real laboratory is at sea," he later told a student. "Buildings are just places to house data." Elsewhere, it was remarked how a pipsqueak lab had provided itself with the largest research ship in the world.

Vema's first scientific voyage, to the Gulf of Mexico in the spring of 1953, was the first to carry a precision depth recorder. "I swore," said Ewing with special passion, "all the years I was at Woods Hole, that when I got control of a ship I'd put on a decent echo sounder." The old echo sounders that he had explored with for seven years revealed an ocean floor that single soundings by line and sinker could not have hinted at, but traced Sixty-Cycle Ridge as well as the Mid-Atlantic Ridge. The PDR had to be damped so it would not super-impose the ship's rise and fall in waves onto the bottom topography. It took very little time to cross abyssal plains with it and settle arguments about their depth and flatness and which way they sloped — away from the submarine canyons that fed them. With the exception of an occasional seamount, there will be, for tens and hundreds of miles on the PDR record of an abyssal plain, nothing large enough to see. The PDR made it abundantly clear that it is only the low places where sands — turbidity current sediments — collect; and that any slight rise is draped in normal abyssal sediment. On the Sigsbee Abyssal Plain in the middle of the Gulf of Mexico, *Vema*-1 discovered, according to J. L. Worzel, more salt domes than all the oil companies had found in ten years. The Sigsbee Knolls are low hills nearly buried by the plain in several thousand fathoms of water. Oil companies paid for the third voyage of *Vema*, which returned to the knolls. *Vema*'s engine failed in the Gulf, and she was sailed back to Palisades.

Like *Atlantis* she was a heavy roller, and hove to she built up resonance until she put her rails under; fifty-degree rolls were no rarity. Under sail she would roll to windward as well as to leeward; sailors said she had lost stability along with her topmasts. Her early cruises were marked by silent comedy confusion at sailing time; some still prevails. She tied up at a desolate pier several miles from Lamont. Data and cores were removed for study. For weeks a stream of cars came from Lamont, taking equipment off for overhaul and bringing it all back. Provisions and spare parts, pipe for coring and storing sediment, were loaded and stowed. Labs were rebuilt and re-wired. Then the excess population of the big house packed itself aboard with its dunnage and last-minute additions to instrumentation. *Vema* moved to the Navy ammunition depot in Raritan Bay to take on explosives for seismic work, and finally sailed. At foreign ports there were interminable delays. *Vema* finally left Spain in 1954 with a new cook; for the last week of the voyage the crew had nothing to eat but sauceless spaghetti, and water was rationed for the last two weeks. During an early Suez crisis she opportunely put in to load explosives at Aden; the British coolly smuggled the TNT through the streets in an unguarded truck and were glad to see the last of it. Another cruise took on explosives at Casablanca during anti-French manifestations, and, half a day out to sea, threw the well-marked packing crates overboard; the police soon began noticing the crates in the houses of thrifty natives, and prepared for imminent revolution. Lamont ultimately received a stiff note from the State Department. The early cruises were limited, by restricted research contracts, to two or three months, and *Vema* could not reach much of the ocean. It began to rankle with Ewing that over five percent of each cruise was spent steaming over the same ground off of New York. Before Washington wrote contracts for as long as a year, Lamont patched them together and took longer cruises. *Vema* got to Capetown. Rather fondly, old Lamonters recall that Ewing had to fly specially to Washington for money to bring her back. The International Geophysical Year brought the first support for large ventures, and *Vema* got to the Indian Ocean on her fourteenth voyage in 1957. *Vema*-18, starting in 1961, went around the world (*Vema* was the first American research ship to do so), a course generally followed since both by *Vema* and by *Robert D. Conrad*, a new Navy-built vessel that joined *Vema* in 1962. Although Ewing called it one of the themes of his life to get two ships for seismic refraction work, he has yet to feel he could afford to put two of his own ships on the single project. There is too much else to see.

Lamont began to study the water that washes over the rocks and carries the sediments to the bottom. Radioactive dating showed Atlantic bottom water (surface water, chilled near the poles, sinks to the bottom and flows toward the equator) has been out of contact with the air for some six hundred years. A new biology department began to study the life cycles of the little creatures — generically referred to at Lamont as bugs — whose remains make up most of the sediments. Off Peru, a cruise dredged up *Neopilina (Vema) ewingi*, a new genus of mollusc, and several other new genera. The order Neopilina are truly living fossils, dating back to the Devonian period, and so primitive that all other molluscs could well be descended from them. But, overall, Lamont's investigations showed that the abyss has a population of species which is far younger than the populations of other places. The Neopilina are among the very few that are truly antique. So much for living fossils. The first firm evidence of how deep whales dive emerged in the study of telegraph cables for signs of turbidity currents. The remains of sperm whales are found entangled in cables brought up for repair from depths of over a mile. The whales seem to swim along the bottom at a thousand fathoms with their mouths open to scoop up squid; unlucky ones catch a loop of telegraph cable. From Lamont's coring surveys of the ocean floor, Ewing and William Donn developed a theory that a new ice age is on the way. Studying bottom samples, Ewing decided the Arctic Ocean had been ice-free during ice ages. Melting glaciers now are raising sea level, increasing the flow of warm water through the few quite narrow and shallow channels that lead into the Arctic Ocean. The icecap is already extremely thin during the summer, the ocean warm enough that, once open, it would not refreeze. The evaporation from an open ocean would enormously increase the snowfall around it, probably starting the new ice age. There were persistent rumors around Sneden's Landing in the late nineteen fifties that Maurice Ewing was buying land in the Sahara.

There was invariably more research in mind than money in hand. In the decade after Ewing came, Columbia's geology department staged over one hundred deepwater voyages. Lamont published some six hundred papers in its first fifteen years, of which Ewing was an author of one quarter and first author of fifty-five. In the succeeding ten years there have been fourteen hundred more Lamont publications, a sort of printed turbidity current.

On the morning of January 13, 1954, *Vema*, starting her third voyage, was running off before a nasty winter gale between Cape Hatteras and Bermuda, and pitching and rolling violently. The wind made loud noises in the rigging, the gray seas assumed the aspect of broad hard hills and valleys, and on each roll the scuppers caught a piece of the passing sea and sluiced it across the decks. It was early morning, gray and luminescent, and many men were still asleep. Ewing crossed the deck forward to the chartroom to check *Vema*'s position. A few rolls later, the lashings around four fifty-five-gallon drums of lubricating oil gave way. The drums began to rumble back and forth across the deck, battering gear and fittings and threatening to stave in the wooden deckhouses. Ewing, his youngest brother, John, and Charles Wilkie and Michael Brown, the first and second mates, came on deck in a hurry. They captured the drums one by one and got them back in place and secured. They were just stepping back and saying, "Those will never come loose," when a freak wave washed all four men and all four barrels overboard. In seas of any size there are waves traveling different courses and occasionally merging with each other; and the odds are that every thirteenth wave will be twice as large as the average, every thousandth three times as large, every three hundred thousandth four times as large. No one saw which it was that washed over *Vema*. Ewing said afterward that he suddenly became part of an emulsion of men, oil drums and seawater. Water almost covered the ship, and then she rose clear. When Ewing came to the surface, he saw the two mates holding on to a drum, and John swimming toward the ship's log line. He himself had swallowed a lot of water and felt in poor shape. He tried to swim to a drum, failed, and began shedding the heavy clothes that weighed him down. In a letter he wrote to his children afterward, he said he had wondered how long it would take his shoes to reach bottom three miles down, and thought how silly they would look if one of Lamont's deep-sea cameras should photograph them. He heard a man call, "Doc! Help me," but could not see him; all he could see was waves. Then the voice, Wilkie's coughed, choked, and stopped. Wilkie drowned then. Each time Ewing surfaced, the waves knocked him under again and rolled him around, and he swallowed more water. He concentrated on floating, and breathing, and listened to the water bubble and rattle in his lungs at each breath.

Vema was about a mile away. There was some genius in the handling of her then. Her complement was, as now, Nova Scotia seamen and fishermen familiar with trouble at sea; they expected, and received, no chance to try again. The captain had brought *Vema* about, but couldn't see the men in the water for the seas. The sailing master took the wheel while the captain climbed to the crosstrees and signaled where they were. Things larger than men can disappear in even moderately rough water, and it is considered impossible to get a ship close enough to a man to pick him up without running him down — seamanship manuals prescribe dispatching lifeboats, while the men in the water patiently wait. The sailing master, however, had been the captain of *Atlantis* in the thirties when

Ewing first sailed on her, and he had had much practice bringing a ship gently alongside small floating seismographs (and he had signed on for this voyage — from a sailors' retirement home — at Ewing's insistence). *Vema*, her present master says, handles like a sports car. She stopped just upwind of John Ewing, and he was thrown a line and pulled aboard. The log line had moved so fast it burned his hands, but he had found a ladder in the water, and floated on it in what he said afterwards was comfort. His brother thought, when he saw the ship stop, that her steering had failed again. It had the day before. *Vema* started again, but turned away from him — he supposed in order to come up on the other tack. He was not doing well, and wondered if he would last until she reached him. He couldn't swim, and couldn't hold his breath and float. For some moments he blacked out. To his children he wrote the next day: "I guess you'd think that a person would be pretty much alone out there at a time like that. I wasn't alone a bit. It seemed as though all the good people I love and who love me were there, and were encouraging me. Then they all went away and just you children were there, and it seemed that I needed to come and do something for my children. It seemed that all of you were about to drown, and I had to keep swimming to save you. Then only little Maggie was there. I couldn't see Maggie, but I could hear her. She was calling just the way she calls down the stairs when she hears my voice when I come home at night."

Then a loud voice said: "Doc, I could hold on to this barrel easier if you'd take hold of the other end." Mike Brown, the second mate, was there, pushing one of the oil drums before him. They both held on and their weights balanced it. The ship was coming toward them and a line was heaved. Brown caught it and, still holding his end of the drum, let the men on board pull them to the ship's side. *Vema*'s rails were going under with each roll. As she came down, Mike Brown put his arm over the rail and went up with her. The same roll pushed Ewing under. "I saw a rope right beside me just as I went down, and I got hold of it. I thought I'd never come up, but when I did I still had hold of the rope. And then on *Vema's* next roll the men caught me by the arms. I don't remember anything more for a long while, until I woke up in a bunk under a pile of blankets.

"Now it's the next day and I'm awfully thankful to be alive. I have learned that the steering gear broke down just as the man threw the heaving line to us. If it had happened sooner I could not have been saved." Or if *Vema* had been slower or if the man with the line had missed, or Mike Brown lost his grip. *Vema* drifted for an hour while the steering was fixed. Ewing had a concussion and his left side was partly paralyzed; next day he was put in a hospital at Bermuda, but he was back aboard before the end of the voyage. John Ewing had a battered leg which soon healed. Mike Brown took a shower and stood his watch.

Eight

Rift and Ridge

A science scarcely can be said to exist before its material, whether it be species or elements or geologic and geographic features, has been organized and classified. By the coming of *Vema*, Lamont had started on a sea-bottom map of the North Atlantic; and one of the first fruits of examining echo sounder profiles was another basic, brutal fact like the existence of the oceanic crust. Ewing gave Bruce Heezen, a graduate student, charge of translating the existing soundings into a chart of the bottom. Heezen is a round-cheeked and round-eyed, and generally slightly disheveled-looking scholar. As an undergraduate in Iowa he was converted from fossils to the ocean floor by a lecture Ewing gave at his college. Now associate professor of geology at Columbia as well as researcher at Lamont, he has

A regiment of unearthly looking sea cucumbers — which are animals —
bears down on the remains of a fish. The Aleutian Trench, 23,976 feet.

An acorn worm and its trail, a hieroglyphic that puzzled abyssal photog-
raphers for many years. East wall of the Kermadec Trench, 16,000 feet.

Sponges and a branched coral growing on bare, igneous rocks of the western rift mountains. North Atlantic Ocean.

been an imaginative and prolific investigator. "You've got to be kind of outrageous," says he, "or scholars use your ideas and ignore you. If they're trying to refute you, they don't mind crediting you — and you stimulate them to do something." In this spirit, perhaps, Heezen has maintained for nearly fifteen years, as a way of accounting for certain characteristics of the ocean bottom, that the earth is expanding. The chart of the North Atlantic ocean floor that Heezen produced after six years — charts of the South Atlantic, Indian and Southwest Pacific oceans have followed, while the North Pacific has been mapped by Scripps — is not the conventional contour map but a sketch map of the features themselves, a form called a physiographic diagram. Heezen says there were two reasons for the choice of procedure. From 1952 until 1962 the United States Navy classified all soundings from the deep ocean, for national security; oceanographers were not permitted to publish the precise locations of their work. Moreover, there was not enough data to make contours. Poetic truth was decided on. In 1953, there were only six topographic profiles across the North Atlantic, Heezen says — five made during the Mid-Atlantic Ridge voyages of 1947 and 1948, and one made from the *Kevin Moran* in 1952. Only west of Bermuda were data plentiful. "You can't imagine how primitive our knowledge was then," says Heezen. "Just the problem of what was there on the bottom occupied us completely. We were in a state so primitive that other scientists couldn't even understand how this could be science." In the next few years there was an explosive increase in echo sounder crossings of the ocean. There were several dozen, all essentially similar, of the North Atlantic by 1959.

In that year the Geological Society of America published *The Floors of the Oceans I. The North Atlantic*, by Bruce C. Heezen, Marie Tharp, and Maurice Ewing, a small and classic book describing and depicting for the first time the floor of an ocean. In the accompanying map are the abyssal plains and abyssal hills, the gorges of submarine canyons, the scarps of the continental rise, and the massed peaks of the Mid-Atlantic

Profiler record across the North Atlantic, Dakar to New York, made during *Vema-22*, in 1966. From the African coast, the sediments thicken steadily

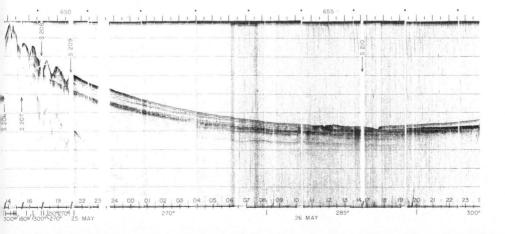

Ridge. The Atlantic, South as well as North, is particularly symmetrical among oceans, as symmetrical as some of the small seas. But the same regions are found in all oceans, and are governed by the same processes. Though not all of these processes were understood when *The Floors of the Oceans* was written, the description has held good, for the Atlantic and for other oceans; the book has no sequel. (The ocean floor between North America and Hawaii was described in 1964 by H. W. Menard of Scripps, in *The Marine Geology of the Pacific*.) At the center of the Atlantic lies the Ridge, occupying the central third of the ocean, a width of a thousand to twelve hundred miles. The borders are the continental margins. Between the margins and the Ridge is the ocean-basin floor, with the great stretches of abyssal plain.

It is a world of dark and shadows, and larger than the dry land. The imagination constructs it from the scattered echo sounder records. Start on the ocean floor off the eastern seaboard, between New York and Bermuda, where there are many crossings. A plain stretches seaward a thousand miles without a blemish as high as a man. Landward, the continental slope looms above the scene, a bare cliff a mile and a half high. It stretches both north and south for thousands of miles. Its face is scored and fluted by gulches and ravines, like the side

down the continental slope; on the platform of the Cape Verde Islands they are ponded thickly between peaks and rises.

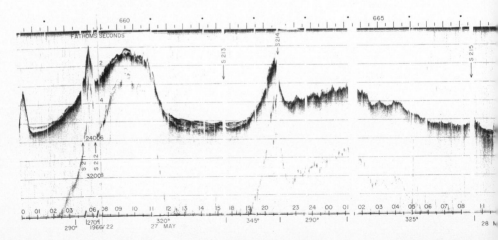

of a dry washout many times magnified. The Hudson Canyon is the largest of these on this coast, deep and wide as much of the Grand Canyon, but to the north and south are dozens of gorges a quarter mile deep. The continental slopes are the highest and steepest, and by far the longest, cliffs in the world. The east scarp of the Sierras, rising precipitously out of the Nevada desert, is like it in appearance, but smaller. Slant and height about match the north face of the Eiger. Off Florida, at the edge of the Blake plateau the bottom drops so sharply that it disappears from the echogram, which picks it up again several miles deeper, several miles out to sea. As Ewing showed in 1935, the continental slope is not structurally the edge of the continent, but eroded sediments and sedimentary rock lying on the gently dipping basement. The top of the sediments is the continental shelf, which is continuous with the sediments of the coastal plains south of New England and has, during ice ages, been dry land itself. At the foot of the continental slope is a gentler grade called the continental rise, the combined deltas of all the submarine canyons and turbidity currents. This shallow sloping plain drops eighteen feet a mile for three hundred miles to the even flatter abyssal plain. Beneath the sediments of shelf, slope, and rise, the transition from continental crust to oceanic crust is made, the

Seaward of the Cape Verdes, a rough basement may be seen through the sediments for the first time. Several crisp layers in the sediments closely

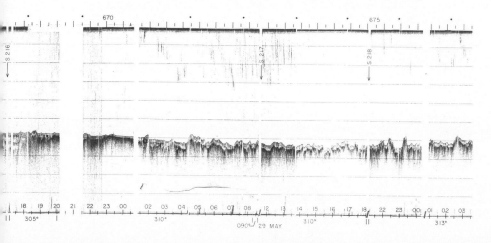

mantle getting shallower as the ocean gets deeper, until there are only four or five kilometers of rock between them. This continental margin is typical of the Atlantic, but in the Caribbean there is a second kind, common in the Pacific and, in appearance, more spectacular. Off Puerto Rico, the continent drops twenty-one thousand feet into the Puerto Rico Trench. This narrow cleft extends from eastern Cuba twelve hundred miles to the Windward Islands; but, compared to the trenches that almost completely ring the Pacific, it is a small one. At the bottom of the trench are two abyssal plains, the deeper, at 27,510 feet, over a hundred and fifty miles long. The outer wall of the Trench is only nine thousand feet high, rising to the crest of a ridge that slopes gradually seaward to the ocean basin floor and the abyssal plain.

Down fifteen thousand feet and more, the floor of the ocean basin stretches off toward the Ridge. Basins lie symmetrically about the Mid-Atlantic Ridge, but are irregular in shape and internally asymmetrical. Off the English Channel and Newfoundland it is only eighty miles from the foot of the continental rise to the Ridge; off Hatteras and Morocco, it is a thousand. Abyssal plains extend seaward from the continent and from mid-ocean canyons and channels; tracts of abyssal hills push into the plain landwards from the Mid-Atlantic Ridge. There

follow the outlines of the basement. (Vertical white spaces are intervals when the profiler stopped working.)

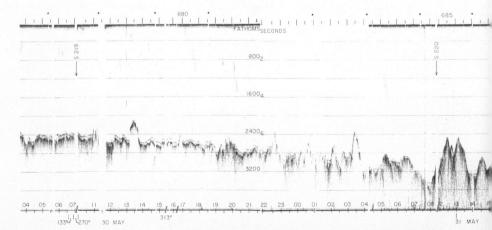

are scattered and clustered seamounts in all basins, and miscellaneous rises. Placid currents whiffle through them, leaving ripples in the sediments of the floor. Minute burrowers eat their way through the sediments, leaving them churned and mottled like marbles. In a few basins there is no circulation at the bottom, and the water is stagnant and sulfurous. On stretches of plain where the sediments collect slowly, manganese nodules are thick as windfall apples, and often larger. They contain relatively pure amounts of manganese, iron, copper, and other elements of salt water that slowly accrete around a shell or a shark's tooth. If anyone can harvest them commercially — several ships already are trying to — they are a lode of fabulous wealth, sufficient copper, nickel, aluminum and manganese in the Pacific alone to satisfy world demands for several thousands of years. Only occasionally since the end of the last ice age, off certain rivers and after earthquakes, turbidity currents thunder through, burying an entire region and leaving the bottom temporarily lifeless and still; the buried microscopic corpses become oil. (Traces of oil only a thousand years old have been found in the sediments.) From time to time, when the cover on the steep slopes of seamounts or hills becomes too heavy, the sediments slump and slide downhill. A great swell, the Bermuda Rise, with numerous

The east flank of the Ridge rises gradually to a peak of 900 fathoms; sediment is progressively thinner, and for 200 miles over the crest it is too

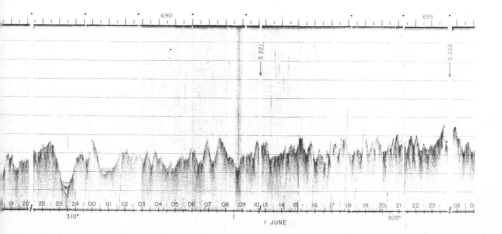

volcanic peaks, occupies the center of the North American Basin, nearly isolated in abyssal plains. Southward is the Nares Plain, named after the captain of the *Challenger*, westward the Hatteras, and north and northeast the Sohm, largest of the plains. The plains, though flat, are ever so slightly tilted — two or three feet per mile — away from their sources. They are at different levels, joined by short necks, picturesquely called abyssal gaps, through which one spills over into another — the Hatteras Plain into the Nares through Vema Gap, the Biscay into the Iberia through Theta Gap. The Sohm Plain has spread like a great sea anemone from the mouth of the Hudson Canyon to the foothills of the Ridge; its broad stalk laps the abyssal hills to the east, and the Bermuda Rise to the west, as far south as the latitude of Florida. Another vast plain lies off Africa, from north of Madeira to well south of the Cape Verde Islands. A sinuous chain of narrow plains flanks Europe. A few huge volcanoes pierce the plain off New England. The single peaks — the largest rises twelve thousand feet out of the plain — form a chain, Kelvin seamounts, crossing the plain southeasterly for six hundred miles. Near the Ridge are the Corner Rise and Seamounts. Off Gibraltar, another group of large peaks, Atlantis, Plato and Meteor, is arranged in the same direction. (The Pacific has more seamounts, but many are low,

thin to see. Between peaks, sediments lie ponded, and in places the movement of the Ridge has tilted them, and faults can be seen.

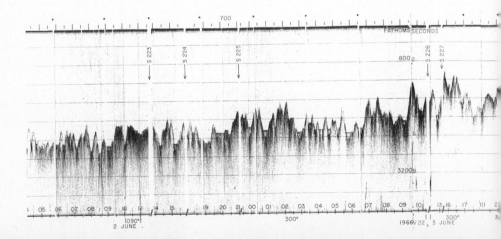

broad, table mountains, called guyots by Harry H. Hess of Princeton, who discovered them during World War II.) The tops of the Kelvin seamounts (a couple are flat, eroded by waves in some former time) lie in five hundred fathoms, but seamounts are often much shallower, shading as a species imperceptibly into islands. The Vema Seamount off South Africa rises to less than twenty fathoms (some seamounts in the Pacific rise to ten) and has been recognized as both a commercial source of rock lobster tails and a danger to ships in heavy seas. Most seamounts, however, are not dangerously shallow, and have yet to make the navigation charts. Once, in the middle of the South Atlantic ocean, a freighter came miles out of her way to inspect *Vema*, which was (quite properly) displaying the at-anchor signal from her masthead, since she was at anchor, on a seamount.

Fed by sediment from the continent, the abyssal plains are still spilling from their borders into new ground, chiefly the abyssal hills. The surface which the plains have already covered, wrote Heezen, Tharp, and Ewing, should also be abyssal hills. It is now possible with an instrument called a seismic profiler to see, through a kilometer of sediment, the rough, hilly bottom under the plains. (The profiler employs much the same process that Ewing used on *Atlantis* in 1947, but done

From its crest, the Ridge subsides to the west in jagged peaks mirroring its rise from the east. The rift valley is just west of core no. S230.

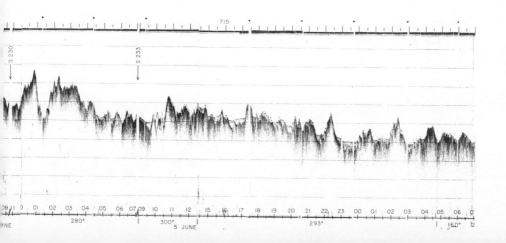

twice a minute instead of twice an hour.) As for the abyssal hills, they "are identical to the small hills which rise from the steps of the Mid-Atlantic Ridge, and probably of the same origin . . . [they] can be distinguished topographically only by the contrast in level." Thus, while turbidity currents make the basins a third oceanic province, there are, structurally and genetically, only two — the continental margin and the Ridge.

The most far-reaching discovery to come out of the oceans emerged during the patient collating and mapping, ashore, of shipboard data. As with many significant discoveries, the manner of its emergence was somewhat fortuitous, even trivial. Marie Tharp was draughtsman for Heezen and others at Lamont in 1952; Heezen says he promptly solved conflicts in her schedule by eliminating all work but his own. They have worked together ever since. "Marie's job for me was to decide what a structure was — whether a rise in the echo soundings represented a hill or something longer like a ridge — and to map it. In three of the transatlantic profiles she noticed an unmistakable notch in the Mid-Atlantic Ridge, and she decided they were a continuous rift valley and told me. I discounted it as girl talk and didn't believe it for a year." By then Heezen had a contract to study cable routes and potential

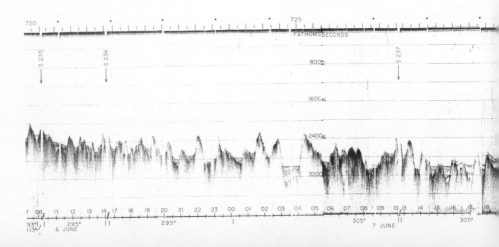

sources of cable failure like earthquakes. The Mid-Atlantic Ridge had been known since the twenties to be the site of some earthquakes; while this was true enough, Ewing and Heezen found they could describe the situation more exactly, and far more interestingly. Neither the locations of earthquakes nor the location of the Ridge had been known in any but the vaguest way. Ewing and Frank Press were then gathering records of seismic surface waves; and the only source from which earthquake waves could reach Lamont after crossing an ocean — but without also crossing part of a continent — was the Mid-Atlantic Ridge. For his cable study, Heezen had another draughtsman plot the new earthquake centers on the same chart Miss Tharp was putting topography on. They fell in a narrow band, not all over the Ridge. "And it was right in the center, where Marie said there was a rift valley — only then did I believe what Marie was trying to tell me," says Heezen, deprecatingly. "Probably anyone with six profiles would have found the rift valley, but in those days no one else had six profiles. It was the combination of the profiles and the earthquake records that made it certain, and you had to be in possession of the data."

After filling the motionless abyss with the tumult of turbidity currents, Lamont was on the track of another, larger,

The west flank of the Ridge subsides into foothills, and these in turn are lost in an abyssal plain at 2,900 fathoms. The basement deepens beneath

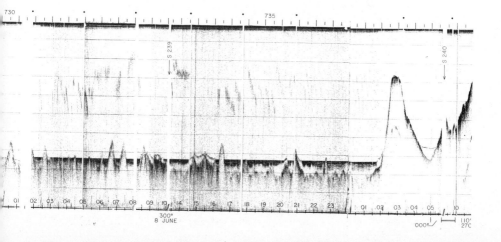

kind of action. Earthquakes at the rift meant that it was currently active, a fact of the highest magnitude and tremendous implication. But they meant more than that to Ewing. He and Heezen began pursuing implications. Ewing said: "It was immediately striking that the earthquakes were all within seventy miles on either side of the rift, and that seventy miles was the margin of error in our means of locating them. Probably all the earthquakes were in the rift valley and the width of the earthquake zone was the error. This has proved to be true. We began to collect earthquake data from every source to add to the records we were making ourselves, and we established seismographs in the southern hemisphere where stations were few and far between — IGY, the International Geophysical Year, came along right then, and we were able to get some support. The books showed a broad band of mid-Atlantic earthquakes from the Arctic to the Antarctic. There were broad bands of mid-ocean earthquakes through all the oceans. We started relocating those old earthquakes by more accurate methods than had been used before. It was inconceivable that *Vema* could ever survey the entire length of the Mid-Atlantic Ridge to see if the rift valley continued, or try to follow it into other oceans should it join a few pieces of high country that we knew about, like the Carlsberg Ridge, south of Arabia, or

the plain, which is interrupted by the northern edge of the Bermuda Rise and some of the Kelvin seamounts. Sediments have filled the valley bottoms between the seamounts.

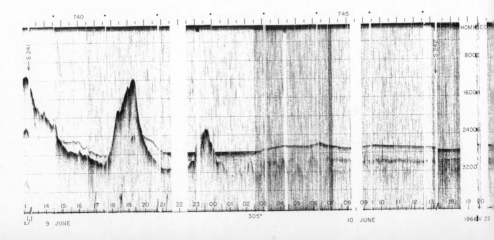

the Albatross Plateau, west of Central America." But the survey could be done with earthquakes, and in a hurry. As far as the Mid-Atlantic Ridge had been seen, the earthquake belt followed faithfully, narrowly, at the center, where the rift valley presumably was. Then, toward the Arctic and Antarctic, the Ridge vanished into areas where there were no soundings. The earthquakes — narrowly — continued. Wherever they came to soundings, there was a ridge. The belt went through the Carlsberg Ridge in the Indian Ocean and the Albatross Plateau in the Pacific; it came ashore into the African Rift Valleys and the San Andreas Fault. By 1956, Ewing and Heezen were ready to present some of their ideas to the American Geophysical Union in a talk called "Mid-Atlantic Seismic Belt" and in a paper with the unpromising title "Some Problems in Antarctic Submarine Geology" (much of the earthquake belt is in the southern ocean, and orgies of international research and cooperation were then afoot for IGY). They had a narrow belt of earthquakes that was forty thousand miles long, reaching through all the oceans and around the world, always (it was to be assumed) in the company of a ridge and a rift valley. It was an astonishing thought, a vast jump in scale from submarine earthquakes or ridges, which separately had never been in question.

The abyssal plain continues deepening toward the continent, the surface broken by an occasional seamount. Just above basement is a strong reflection

But, says Heezen, "this idea of a rift valley was so revolting to geologists that some of them from Scripps went right out and, they said, proved us wrong." Many geologists had thought the idea of rift valleys pretty revolting even before someone came along saying there was one running all around the world. In East Africa there are long, narrow, cliff-sided valleys — including Lake Tanganyika, whose flanks fall over eight thousand feet to below sea level — running from Ethiopia to Rhodesia. These rift valleys, on first geological inspection, appeared to be places where the earth had been pulled apart and the rents filled with lava and rubble. Some geologists, however, even without actually seeing the rifts, found the thought of any real widening in the earth unaccountable and unacceptable. Compression was the force whose signs geologists found most displayed about the continents, especially in folded mountains, and there were ingenious scholarly attempts to show how local tension and rifting in East Africa could be produced by large-scale compression. A worldwide rift was less than wanted. In their eagerness to dispose of a rift, however, scientists confirmed the existence of a worldwide ridge. The Scripps expedition found a ridge in a part of the ocean where Ewing and Heezen's narrow line of earthquakes predicted it, and where no ridge of any kind had been ob-

from Layer A (Chapter 10). Basement disappears as the bottom rises toward the continental shelf. Before reaching the edge of the shelf the ship turned to cross the Hudson Canyon.

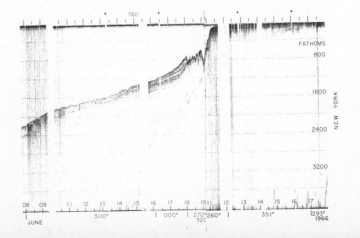

served or suspected. The ridge was there, though not "magnifi-
cently there" — as Heezen remarked after *Vema* went into the
Indian Ocean and made eight crossings of an uncharted ridge
(and rift) as ragged as the Mid-Atlantic one.

The name Mid-Ocean Ridge was quickly applied to the
newly discovered — or newly expanded — worldwide feature,
and geologists now call the whole "the Ridge," while specify-
ing its segments Mid-Atlantic, Mid-Indian Ocean, and so on.
In the Pacific, the Mid-Ocean Ridge is low and broad, with
shallow relief and few high peaks. As the scientists from
Scripps were careful to point out, there is no rift. Lamont had
had the luck to start its exploring where there is one. The
Pacific has altogether less sediment cover than the Atlantic,
and the low Ridge there is not in as striking contrast to its
surroundings as is the Atlantic Ridge; expeditions had passed
right over parts of it without thinking much of it. There was
room to argue, after the finding of the Chile Rise branch of the
Ridge in 1958, that it and other rises in the Pacific were different structures from the more commanding Mid-Atlantic Ridge;
and some scientists did, though the continuous earthquake
belt was a brutal fact that was hard to evade. But the opposition was not great or very determined. The Mid-Ocean Ridge,
and rift, was one of those ideas for which so much evidence
had been lying about in different places that, when someone
finally put it all together, there was not much ground for ob-
jection. And the more ships went out, the more it appeared
that the Ridge — high ridge or low ridge — is continuous from
ocean to ocean like the earthquake belt. The idea of a ridge,
however, was something wholly new on earth — one geologist
called it, without overstatement, the most exciting discovery in
twenty years — and no one yet knew just what it meant.
(Finding that out — only in the last few years — left scientists
beggared of superlatives.) In succeeding years, when money

Abyssal landscape. In the distance, a guyot with the characteristic low, flat shape of an undersea volcano. While still below sea level, the Afar depression of Ethiopia is a part of the Red Sea floor that has dried out. International teams of geologists under Haroun Tazieff have explored it in the last few years.

could be got, ships were sent out, and every time one crossed the earthquake zone, there, also, was the Ridge. The mapping continues; the Ridge has been crossed a few thousand times, but there are many wide spaces between crossings that it would be interesting to fill.

The Mid-Atlantic section of the Mid-Ocean Ridge rises nearly two miles above the adjacent plains. It fills the central third of the ocean basin. Lofty peaks at its crest soar ten thousand feet above the floor of the rift valley without intervening shoulders or steps. The mountains are harsh and unweathered, draped in only a light sprinkle of sediment. While the Ridge gets steadily lower toward its flanks, the peaks there are just as sharp and ragged as at the crest. It is like the inside of a shark's mouth. The rocks are fresh and unworn; coring, dredging, and photography commonly encounter bare rock. Dividing the Atlantic Ridge down the middle, the rift valley falls, on an average, six thousand feet beneath the central peaks; and it is eight to thirty miles from one wall to the other. It is a gulf into which the most generous sixty miles of the Grand Canyon could easily fit. This topography continues along nine thousand miles, from the top of the North Atlantic to the bottom of the South. The earthquakes run north from the Atlantic through Iceland, the Norwegian Sea and the Arctic Ocean — virgin territory in the 1950s — and peter out in the Siberian continental margin well above the Arctic circle north of the Lena River delta; the Ridge, too, peters out into a deep trough in the Siberian continental slope, which was first identified as a submarine canyon. Southward through the Atlantic, the earthquake zone precisely bisects the ocean, following the switchbacks of both coastlines at the equator. It turns east below the Cape of Good Hope and runs northeast to mid-Indian Ocean beyond Mauritius, where it divides. One branch goes northwest into the Gulf of Aden, and thence, splitting again, north into the Red Sea, Gulf of Suez, Dead Sea and the Jordan Valley, and south through an area of recently emerged sea bottom into the rift valleys of East Africa. The remaining branch crosses the Indian Ocean southeastward (here the Ridge becomes less rugged, and the rift disappears), rounds Australia nearly halfway to Antarctica, and crosses the Pacific in a grand sweep to the Northeast. Near the Americas, the Ridge is called the East Pacific Rise. West of Chile, a branch goes toward Patagonia. The Rise passes just west of the Galápagos Islands, gradually closing with the coast, and enters the Gulf of California. The earthquake zone carries on into the seismically active western part of the continent. A branch goes through the Salton Sea and along the San Andreas Fault, and, with the Ridge, which appears again at sea off Cape Mendocino in

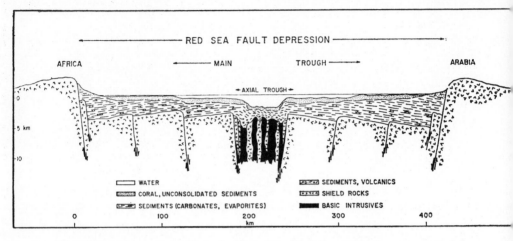

The shape of a young rift and ridge (the Red Sea Ridge) as suggested by geophysical measurements.

The ridge and rift in the Afar depression. Faulting has given the flanks a steplike outline, as in drawing above.

northern California, disappears into Alaska. What goes on in Nevada, the hot springs of Wyoming, the Colorado Plateau, the Rocky Mountain Trench, and other recently uplifted or depressed areas is unclear; the connection with the Ridge is vague. The Ridge may be underneath, much muffled by the great thickness of the continent, preparing rifts like those of Kenya; or the continent may have largely smothered the Ridge. Geology has several versions of reality.

The Ridge rises above water at Easter Island, Rodrigues, and Tristan da Cunha; in ten years, it has added new islands to the Azores and Iceland. Throughout its forty thousand miles it is volcanically as well as seismically active, and the locus of an abnormally high flow of heat from the interior of the earth; magnetometers also register a large disturbance, suggesting lava flows. Iceland is not an emerged peak like the other islands: here it is more as though the whole width of the Ridge, narrowing into the Norwegian Sea, and all its peaks were fused into one blob. Ridge mountains abut the island north and south; the seismic zone runs right through it. There is a central lowland of a kind geologists call a graben, flanked by faults along which it has slid down. This is a place of fissures, hot springs, and lava flows; it is said to be getting wider at a rate of three and a half centimeters a year — but it is the business of rift valleys to get wider. The Ridge is the largest single geologic or geographic feature on earth, equal in area to all the dry land of the continents. "Imagine," said Ewing, "millions of square miles of a tangled jumble of massive peaks, sawtoothed ridges, earthquake-shattered cliffs, valleys, lava formations of every conceivable shape — that is the Mid-Ocean Ridge." During the sixties, the tangled jumble began to make sense.

Nine

Diluvialists and Other Fauna

"Geology" is a coinage of the late eighteenth century, when Dr. James Hutton, a Scot and a physician who spent all his time diagnosing rocks, wrote: "No powers are to be employed that are not natural to the globe, no action admitted of except that of which we know the principle, and no extraordinary events to be alleged in order to explain a common appearance . . . chaos and confusion are not to be introduced into the order of Nature, because certain things appear to our practical views as being in some disorder. . . ." Hutton's was a new view of the earth — though men in every age have believed their ideas to be natural. The urge to disable a fact with revealed truths has been strong in geology. Geologists believe that, unlike earlier doctrines, Hutton's, which came to be called uniformitarian-

Devonian gluttony.

ism, is true; and probably they are right, since they often have yielded to the temptation not to abide by it.

In the beginning, the world was believed created in six days, which James Ussher, archbishop of Armagh and primate of Ireland, established as having commenced at the civilized hour of 9 A.M., Sunday the twenty-third of October, 4004 B.C. God created everything and rested; He has not revised his work. He created canyons, sandstone, crumpled beds of rock, beaches far above sea level, limestone caves, buried skulls, and, best of all, fossils. By the seventh day He had to rest, weak with divine laughter. Fossils were a vexation to scholars and theologians. They were stones, like those wept by unicorns or grown on the heads of toads. Though many of them closely resembled certain living things, what the "figured stones" were in reality was "sports of nature." Albertus Magnus described a fossil tree branch bearing fossil birds in a fossil nest. Fossils could not have been alive, because God created the beasts of the earth on the sixth day, after He created the dry land.

Therefore, there was no way imaginable (save a heresy) for the remains of the beasts of the earth, much less of the fish of the sea, to have become part of the dry land. It was suggested that figured stones were made by God that man might the more admire his skill, or by the devil, that man might fall into the sin of thinking them animal relics, or, most ingeniously, that they were the remains of trial creations which God aborted after He looked on what He made and saw that it was not good enough. The most reasoned and sober opinion of the seventeenth century was *"lapides sui generis*, naturally produced by some extraordinary virtue, latent in the earth, or quarries where they are found."

In the early eighteenth century, a Frenchman went to the Inquisition for arguing an organic origin of fossils, but orthodoxy generally had fallen back on the strategic compromise of the Flood. There are strong signs that the earth has been revised since the creation, and Noah's flood was the only possible biblical cause. Diluvialists, as they were called, counted eminent scholars in their fold until the middle of the last century. The Flood, which Ussher fixed at 2349 B.C., explained just about everything equally well — fossils, canyons, rocks formed under water — and permitted some rocks to be older than others by a few days. Even when such matters were being explained in other ways, the Flood was kept from well-earned retirement as an alternative to believing in continental glaciers. Only in 1846 did the Reverend James Buckland, no country cleric but the president of the British Royal Society and a sincere scholar, recant the Flood before his fellows and adopt the glacial theory; American scholars persisted some years more. Before it evaporated, diluvialism was fruitful and multiplied; a dozen or so "Noachian" floods were appealed to by at least one authority — there was much for one to explain — keeping the ark market brisk.

For a while the civilized world believed in neptunism. In *Faust,* angels speak as with the tongues of neptunists, while the devil sounds like one of the discordant plutonists. Neptunism was grand, celestial, serene, like the spheres; plutonism was mean, smoky, and disjointed. In neptunism, religious dogma was replaced with scientific reason, according to Abraham Gottlob Werner, whose theory it was. Werner was professor of mineralogy at Freiburg's school of mining, a post he attained as quite a young man. His students were inspired as by a prophet, and occupied chairs at universities throughout Europe; his views dominated a generation. Werner thought all rocks — like granite, coal, limestone, and basalt — had been deposited in a former global ocean deeper than the highest surviving mountain; the ocean receded slowly, into the waters under the earth, and rock was precipitated out, one kind after another. He was challenged over basalt, so the opposition (who insisted that basalt was not aqueous but igneous) were called plutonists. "I hold that no basalt is volcanic," Werner wrote, "but that all these rocks . . . are of aqueous origin." Plutonists described basalts in the Auvergne that preserve all the appearances of having once flowed, like lavas. Werner replied that beds of basalt must once have lain over beds of burning coal — the cause of volcanoes — and been melted by the fire. The sensible Scot James Hutton was a plutonist, though he held that all lava is extruded underground, in dikes or sills like the Hudson Palisades, and none by volcanoes. But his neat demonstration of granites that have baked the adjacent sedimentary rocks was one of the early and classic proofs in geology. Unlike their master, Werner's followers were vociferous about their views, and one, shown Hutton's evidence for molten granite, "did not attempt to explain the facts before him on any principle of his own . . . but he burst out into the strongest expressions of contemptuous surprise that a theory of

157

the earth should be founded on such small and trivial appearances! He had been accustomed, he said, to look at Nature in her grandest aspects . . . and he could not conceive how opinions thus formed could be shaken by such minute irregularities as those which had just been shown to him. [His hosts] were confounded; and, if we recollect rightly, the weight of an acre of fiorin and the number of bullocks it would feed formed the remaining subjects of conversation." After decades of argument, Wernerism was demolished when three leading disciples finally visited the Auvergne, around 1805; they adopted the plutonist interpretation. Geologists became inveterate travelers. (Werner said he would have to see the Auvergne for himself and never went. In his later years he stopped reading professional journals, writing letters, even opening his mail; for months he was unaware of his election to the Academie Française.)

William Smith is a relief after the abstractions and contentiousness of the learned. He was a surveyor, a country man not an academic, who tramped and drove about the land all his life. He advanced no theory of how or when the earth was made or by whom, but from small and trivial appearances he taught himself how it is put together. He was said to be able to tell, just from a glance at the surface, what lay below. Ultimately he impressed the learned even more than he had his customers. Some of them persuaded him to make a map of the English rocks, the first geologic map, and it was said he almost wrote himself out of a living. Smith was one of the few innovators who aroused little opposition; indeed, he seems to have been more reluctant to take time from surveying than others were to hear or read him. He acquired a nickname, "Strata," before he died — one wonders who employed it, and how — and was called "the father of English geology" by the Royal Geological Society. Smith understood as no one had before him that rocks which dip under the surface in one place reappear in another without change, and that the order of the layers here is the same there. All geological fieldwork, scholarly or commercial, is done with that assumption.

Smith saw also the basis of historical geology: that each of the regular succession of rocks is characterized by unique fossils — with which he had as much familiarity as with strata. A Reverend Mr. Benjamin Richardson, of Bath, who persuaded Smith to begin publishing his observations and who was himself a collector of fossils, was much impressed that Smith could tell him more about his own collection and where it came from than he knew himself. Many a cultivated gentleman amused himself picking fossils on his rambles and regarding them over a glass of port, without always much idea of learning from them — they were curiosities, perhaps antediluvians. The Reverend Mr. Richardson reports on the emergence of one of the great ideas of science: Mr. Richardson and a friend, the Reverend Mr. Townsend, "were soon much astonished by proofs of his own collecting, that whatever stratum was found in any part of England, the same remains would be found in it and no others. Mr. Townsend, who had pursued the subject forty or fifty years, and had travelled over the greater part of civilized Europe, declared it perfectly unknown to all his acquaintance, and he believed, to all the rest of the world." One can imagine a cluttered scholar's study, and the outdoorsman, as the two clergymen listen, dividing the good Mr. Richardson's fossils among the homely English strata — Old Red Sandstone, Wenlock Shale, Bath olite . . . — and putting them in their rightful sequence, rather as if he were arranging dominoes for two men who couldn't count. It is so obvious and right when it is pointed out. With just this tool, the history of the world as the rocks have recorded it was understood after not very many years. Geology began over and (very much as it is doing today) erected new theories.

For years there was enormous popular interest in fossils.

The Philosophical Magazine and Journal of 1815 printed an article entitled (text was superfluous) "An Earnest Recommendation to Curious Ladies and Gentlemen Residing or Visiting in the Country, to Examine the Quarries, Cliffs, Steep Banks, Etc., and Collect and Preserve Fossil Shells as Highly Curious Objects in Conchology, and, as most Important Aids in Identifying Strata in Distant Places; on which Knowledge the Progress of Geology in a principal degree if not Entirely Depends." The worlds of the past, which naturalists reconstructed from fossils in their museums, became a source of much popular wonder, as new worlds always have, as the Indies were to the Elizabethans — or television pictures from the moon. From the gypsum quarries of Paris, Baron Georges Cuvier resurrected a host of strange creatures, like pterodactyl. "Is not Cuvier the greatest poet of our century?" wrote Balzac. "Our immortal naturalist has reconstructed worlds from bleached bones. He picks up a piece of gypsum and says to us, 'See!' Suddenly stone turns into animals, the dead come to life, and another world unrolls before our eyes." Fossils impressed Cuvier with the evanescence of life and of whole species, and with the prevalence of catastrophe. Cuvier's doctrine, catastrophism, was a heresy that the priests of uniformitarianism, despite triumphs of natural explication, could not stamp out for decades. In Siberia were frozen mammoths, apparently alive one moment, dead the next, and encased in ice before they could decompose; creatures vanished from the strata often very suddenly, never to reappear. Cuvier wrote: "Life has often been disturbed by great and terrible events which are everywhere recorded. . . . In vain do we search for present causes sufficient to produce the revolutions and catastrophes the traces of which are exhibited in the earth's crust." The antithesis of that declaration has been the theoretical foundation of geology for the last one hundred and fifty years.

Time on an inhuman scale multiplies the insufficient causes of the present into the revolutions and catastrophes of the past. The present differs from the past chiefly in there being less of it. With sufficient time, the transformations and succession in the rocks and the fossil remains are fathomable instead of miraculous, wrought by homely processes seen every day. The stream as it wears down the hills deposits the material for new hills, perhaps upon the stumps of yet older hills. Dr. Hutton wrote: "We are thus led to see a circulation in the matter of the globe, and a system of beautiful economy in the works of nature. The earth like the body of an animal, is wasted at the same time that it is repaired. . . . The operations by which this world is thus constantly renewed are as evident to the scientific eye as are those in which it is necessarily destroyed." In the rocks, he wrote, is described "a succession of former worlds." The very oldest we find (sedimentary rocks just found in Greenland are nearly four billion years old) are composed of the debris of earlier ones. "We find no sign of a beginning — no prospect of an end."

Known only to a handful of friends in his lifetime, Hutton's ideas after his death became, with Smith's practical observations, the kernel of Sir Charles Lyell's uniformitarianism. It was a great many years before Lyell's *Principles of Geology* could be added to in more than detail; Charles Darwin, just ten years younger than Lyell, eagerly waited publication of the last volumes as the *Beagle* picked her way around the world, and found in Lyell the geological framework for evolution. In the eighteen forties, Henry Murchison and Adam Sedgewick (following — in not always friendly rivalry — the strata between where Smith left off and the Irish Sea) pushed the rock and fossil sequences back to what seemed the beginning, the earliest fossils. The Cambrian, the Ordovician, the Silurian, the Devonian — half the names of the Paleozoic periods — derive from Wales and west England. The fossil-barren strata just below the Cambrian were simply called pre-Cambrian, but eighty-five percent of the earth's history turns out to be pre-Cambrian. The amount of time the new geology demanded was disputed for the remainder of the century and it is common still for geologists to speak of periods of mountain-building as revolutions, though they are known to have occupied tens of millions of years. Lord Kelvin, the greatest physicist of his day, gave judgment that "British popular geology is in direct opposition to the principles of Natural Philosophy." He threw geologists into a dither by warning them absolutely that from the laws of physics "we now have good reason for judging that, unless some new source of heat is found, [the globe solidified from a molten state] not less than twenty nor more than forty million years ago, and probably nearer to twenty than forty." Since geologists and evolutionists already required several hundred million years for their chronologies, and were adding more daily, they were perturbed. "The physicists have been insatiable and inexorable," wrote Sir Archibald Geikie. "As remorseless as Lear's daughters, they have cut down the grant of years by successive slices." In 1895, however, Becquerel was discovering radioactivity; and ten years later Ernest Rutherford politely showed Kelvin's figures to be meaningless. The new source of heat had been found. The geologists' trek into the past met no further impediment.

Geologists could see that rains and rivers wear down the earth, that the very substance of mountains is sediments worn off in a yet earlier time, crumpled, turned to stone, and lifted above the plains. Anyone who drives along a highway cut through the shoulder of a mountain can stop and see as much. The earth has lived four times a thousand million years, through cycle after cycle of growth and decay whose remnants are recent Alps, old Appalachians, and mountains so ancient

Sedimentary rocks — which once lay flat — folded, then eroded into mountains. Idaho.

they are only stumps. But if the earth is like the body of an animal, what quickens it? How is it renewed even as it is wasted? A hundred years ago the Irish geologist and seismologist Robert Mallet proposed that the globe is shrinking. Like Kelvin, he supposed the earth is getting cooler; and he thought the solid crust would have to crumple like the skin of a drying apple, creating mountains. Mallet computed a shrinkage of six hundred miles in the earth's circumference since its formation. Many geologists rejected the theory immediately, but had nothing as satisfying to replace it with. The theory still was at the heart of much geological thinking five or six years ago, though nearly a century earlier Clarence Dutton announced implacably: "It is quantitatively insufficient and qualitatively inapplicable: it is an explanation which explains nothing which we wish to explain."

Expedition explores a peak of the Mid-Ocean Ridge. Surtsey.

"Do you know any geologists?" Ewing said one evening a number of years ago, as he sat at the stern of a ship, watching the light fade over a tropical lagoon. "Annoying fellows. I'm a physicist. Geologists spend their time poking around trying to explain this or that little detail. I keep wanting to say, 'Why don't you try to see what's making it all happen?' It's as if you set a battleship in front of these men in dugout canoes, and after a few years they learned how to turn on lights. We go to the empty areas of the oceans, making as many kinds of measurements as we know how. Once we survey a place, we know what to come back and look for. It sounds hit-and-miss, but with some intelligence and a little luck it's a lot more. We can't look at enough. It's my view that we won't know where the most interesting places are until we've seen all of them. It may sound egoistic to want to survey everything, but that's what we're doing. So we can say to the ocean: it is thus and so. We're tracing such monstrous differences across the ocean floors that some great overriding geologic process must be involved. It makes the work terribly exciting right now. I don't want to explain things to people — I want to understand them. Is there a pattern? I feel every night, when I go to sleep, that surely I'll see it when I wake up."

In 1955 Lamont Observatory published the tenth part — *Newfoundland* — of Ewing's survey of the continental shelf, which had become a series embracing the whole east coast, a definitive and surprising study of the edge of a continent. Coincidentally, in a classic monograph called *North American Geosynclines*, Marshall Kay, a geologist at Columbia University, had just reconstructed the probable appearance of the east coast of the continent in the Paleozoic era. Although later it was somehow crumpled and raised into the Appalachian Mountains, in those days before the upheaval it was very like the coastline today. There was a low coastal plain, like that from New Jersey to Florida, and a shallow continental shelf. The shoreline was in Ohio, however. Offshore, to the east, under the continental shelf and margin of the time — present-day Alabama, Pennsylvania, New Hampshire — were two long troughs in the basement rock — geosynclines — filled with sediments. The sediments later were transformed into the Appalachian Mountains. Their amount was enormous. Although most of the original Appalachians have been eroded away, and the detritus carried to the sea, the thickness of sedimentary rock remaining runs to tens of thousands of feet (the sediments of the younger Rocky Mountain geosyncline, if spread out, would cover the United States a mile and a half deep). Judging from the mountains, the inshore trough was the shallower. The sediments that filled it were lime and mud as might collect at the bottom of a shallow sea; they contain small shells, crinoids, and the remains of other early animals and stemmed plants. (One strange thing about the sediments that formed the Appalachians: many appear to have come from land lying to the eastward — that fabulous "lost continent" of Appalachia.) A ridge of the basement divided the two troughs of sediment. In the outer, deeper one, volcanic islands, or seamounts, deposited ash, and the sediments were conspicuously layered and relatively poor in fossils. The sandstones formed of them, graywackes, make the eastern remnants of the Appalachians quite distinct from the western with their shallow-water limestones and mudstones. The sands of the beds are finely sorted and evenly graded from coarse at the bottom to fine at the top. In 1959, Charles Drake, who had come to Lamont as a graduate student soon after it opened, published a description of the geology of the present east coast and continental margin of North America, based on the surveys that had been made over twenty years. It was his doctoral thesis, which he co-signed with Ewing and George Sutton.

Offshore, from Newfoundland to South Carolina, there are two parallel troughs in the crust, filled with sediments and separated by a ridge of basement rock. Shallow-water sediments, full of fossils, and as much as seventeen thousand feet deep, fill the inner trough under the continental shelf. In the outer trough, under the continental slope and rise, the sediments reach thirty thousand feet deep, and they are layered, graded, sorted, and poor in fossils, like the Paleozoic graywackes of the middle Hudson valley. They are turbidity current deposits. "This sedimentary system is quite comparable to the Appalachian system as restored for early Paleozoic time," Drake wrote. "The sediments of the inner and outer troughs are similar to those of the Appalachian [inner and outer troughs], respectively, and the basement ridge resembles the pre-Cambrian axis which separates the two troughs in the Appalachians. . . . The major process necessary to convert the present continental margin into a mountain system is a process to thicken the crust . . . to continental proportions." Here was a mountain range waiting to be born — an extraordinary and unexpected discovery. But the energy that would transform the waiting raw materials into mountains did not appear to be at hand. What is this energy and why is it absent here and present elsewhere? Great chains of mountains are rising even today, like the Andes, while others, like the Appalachians, have long since grown cold. Why? And why are the ranges long and narrow, instead, perhaps, of being round? (There are relics of ranges dating back billions of years that appear to have been quite different from "modern" ranges like the Appalachians.) Why are there mountains at all? Why are there earthquakes? What is tearing the earth apart, and why in California when not in Connecticut? None of this could geologists explain with any assurance, or agreement, even a few years ago — though they could tell the history of the earth in some detail. The questions are questions of geophysics, the specialty devoted to the earthly energies and their manifestations, to gravity and magnetism and heat and earthquakes and their waves. Curiously, it has been exploration of the floors of the oceans that started supplying the answers.

Ten

To Gondwanaland

SINCE 1956, when Maurice Ewing and Bruce Heezen made their laconic and elegant association between soundings and earthquake data which led to the recognition of the Mid-Ocean Ridge, the nature of the Ridge and its effect on the remainder of the world have been the leading problems for marine geologists and geophysicists. With every new voyage to distant seas, the vast extent of the Ridge became more evident, and it became more obvious that the ocean bottom was not the uniform, motionless, monotonous, almost featureless accumulation of sediments it had been thought to be. On the contrary, something momentous was going on there, and Lamont scientists groped at what it was. A Lamont paper on sediments observed that they were unbelievably thin, compared to what had been expected in a world several thousand million years old. There is an average of only fifteen hundred yards of sediment in the Atlantic, a thousand in the Pacific. And no one, anywhere, had come up with a piece of sediment that was really old. For years it had seemed that if you could only find the right slope there would be stunningly old sediments uncovered and waiting to be cored; but a few samples, seventy to eighty million years old — an eighth the age of trilobites, a sixtieth the age of the earth — were the oldest. "It dawned on us near the end of the fifties," says David Ericson, who wrote on the sediments with Ewing and others in much the same terms, "that something must have happened about a hundred million years ago to renew or drastically reorganize the earth's crust under the oceans." It was a startling, even brazen, suggestion to make about two-thirds of the planet. What was this reorganization? John and Maurice Ewing wrote:

> The formation of the [mid-ocean] ridge requires the addition of great quantities of basalt magma and raises the question of its source. . . . We suggest that a convection current system [in the mantle] has contributed the basalt magma and applied the extensional forces to the crust to produce the axial rift. We asume that all [continental] crustal material was collected into one hemisphere by [an] initial current system. The second current, whose pattern is assumed to persist to the present . . . broke the continental mass into fragments which moved [into] the present pattern.

The moving continents wiped out an old ocean floor and left a fresh one in their wake.

Wandering continents — continental drift — were the widely scorned idea of Alfred Wegener, who published it in 1920. Wegener was not a geologist, but a meteorologist, a fact that later seemed significant to critics; and others before him had thought of mutable continents. Yet it was he who formulated the idea in acceptable geological — uniformitarian rather than diluvial or catastrophist — terms, and who attracted professional discussion. "The first concept of continental drift came to me," he wrote, "as far back as 1910, when considering a map of the world, under the direct impression produced by the congruence of the coastlines on either side of the Atlantic. At first I did not pay attention to the idea because I regarded it as improbable." The next year, however, he saw biological evidence for a land connection between Brazil and Africa, and was sufficiently impressed to begin looking for more. In *The Origin of Continents and Oceans*, he assembled a vast quantity of evidence, of varying quality, and made from it a coherent hypothesis. During the Permian period at the end of the Paleozoic era there was an ice age in the southern hemisphere. Glaciers covered large parts of South America, Africa, Australia, Antarctica, and also India. If the continents were in their present positions, the extent of glaciation was enormous; one authority calculated that there is not enough water in the world to make that much ice. While the southern continents were under ice sheets, the northern ones apparently were tropics with the rain forests that later became rich coal fields; Spitsbergen was like the Amazon Valley. The planet appears to have had one cold side and one hot side, if the continents were in their present positions. Between the advances of the southern ice there grew unique flora, a particular assortment of plants, dominated by the giant fern Glossopteris, which is now preserved in low-grade coal in South America, Antarctica, Africa, Australia, and India. Geologists and paleontologists were puzzled not just that the same plants had appeared on isolated continents but that the same combination of them appeared, when such combinations today rarely persist across as little as five miles of water. Moreover, the same vertical sequence — of low-grade coal with glacial till and other strata

Old rocks "peneplaned" or eroded to virtual flatness. On the left are granitic rocks some 1,900 million years old, on the right sedimentary rocks some 2,800 million years old. The fault along which they abut is 900 feet high.

Seams of anthracite coal in the Antarctic, laid down in the Permian some 250 million years ago. Theron Mountains.

— also persists from continent to continent. It is called the Gondwana sequence, from the place of its first discovery in India. Wegener also was impressed with the evidence of living things, particularly the more homely and less mobile, like a garden snail that is found only in eastern North America and western Europe, suggesting it did not amble round by way of the Bering Strait. Oldish genera of worms are found on both sides of the South Atlantic, younger ones on only one side. Australian mammals also were instructive. Neither marsupials nor their fossils are found in Asia, though they occur in every southern continent — between which they could not migrate today without passing through Asia. Marsupials haven't gotten from Australia to the nearby islands, and Asian mammals only recently began to invade Australia. But there are marsupials in South America, though Buenos Aires is as far from Australia as from the North Pole; and some of the marsupials on each continent have some of the same parasites. Wegener concluded that there was only one, jumbo, continent in the Paleozoic era, and, from the dates of his evidence, that it began to break up during the Mesozoic into the present continents. Antarctica, Australia, India, Africa and South America, like the wedges of a pie, had had a common center near South Africa, which was then the South Pole. The separating continents, Wegener thought, barged their way through the ocean floor like ships in pack ice; at their bows were earthquakes and young, growing mountains, and in their wakes new ocean floor was formed. Indeed, the Pacific, on which the drifting continents encroached, is ringed by earthquakes, volcanoes, and new mountain ranges. Wegener candidly admitted his evidence was circumstantial; he did not know what force moved the continents, and his few suggestions were flimsy. But while he believed he had proved the act had been and was still being committed, there were experts who, between them, contested most of his evidence. Indeed, no one has yet shown that the evidence, and more like it that has been added since, is exclusive. It can be interpreted other ways. And it applies to percentages of species and strata — large numbers, but not, apparently, majorities. Someone with doubtful motives suggested that some day a fossil may be found one half of it in Brazil, one half in Africa. The similar coastlines were called coincidence — surely if a continent *did* move, its coastline would be distorted — and it was said that western Australia makes a nice match with eastern North America, geographically and geologically. Many geologists found Wegener annoying — not worth discussion, but necessary to discuss nonetheless because he developed a following among laymen and southern hemisphere geologists. "Contrary to all available physical and geological evidence," wrote the geologist Walter Bucher, "he blandly postulated that the basalt of the oceans is so weak that it cannot resist deformation even under the action of almost infinitesimal forces. . . . a masterpiece of special pleading." The eminent geophysicist Sir Harold Jeffreys did not even bother to invent his own refutation. In his monumental volume *The Earth* (whose latest edition an irreverent colleague recently reviewed under the heading *An Earth*), Jeffreys proclaimed: "We may apply to the theory the words used by Dutton about the thermal contraction theory. It is qualitatively and quantitatively inapplicable. It is a theory which explains nothing which we wish to explain." During the forties and fifties, according to Bruce Heezen, a young geologist who wanted to get or to keep an academic job in this country could not support continental drift too cautiously. The crystallographer Sir Lawrence Bragg has said that the time he showed a paper of Wegener's to a well-known geologist afforded the only occasion he ever saw a man actually foam at the mouth. A wise man might have found reason for believing

Wegener in the reaction to him, though Wegener really was wrong in many respects. "Altogether the polemics on continental drift are a sobering antidote to human self-confidence," a geologist reflected recently. "Practically all the arguments against it, and many for it . . . now appear to have been fallacious." Ideas are currency in science, but they are paper money, worthless without good coin to back them up; even quacks may have the right idea for the wrong reason, while making an idea seem necessary instead of expendable requires some genius.

When an international oceanographic congress convened in New York City in 1959, a number of the best and most resourceful minds in marine geology were there, including J. Tuzo Wilson of the University of Toronto, one of the most fertile minds in the business, Henry W. Menard of Scripps, who almost single-handedly was discovering and explaining the sea-floor geography of the Pacific; the late Harry H. Hess of Princeton, a geologist of great perception; Sir Edward Bullard of Cambridge University, a geophysicist who had done pioneering work at sea; and Ewing and Heezen (curiously, Bullard, Wilson, Hess and Ewing all had been in the circle around Richard Field at Princeton in the thirties). All were distinguished by an interest in both the earth and the sea; and most believed that the Mid-Ocean Ridge must be the source and cause of some sort of wholesale motion of the earth's crust, like continental drift. What form of motion was quite unknown — except that it was not Wegener's — and the subject of the liveliest curiosity. Too tentative to get more than passing mention in the papers delivered, the question was discussed in the halls. Too much was known about the ocean floor — this, indeed, was one reason for having a congress — for geologists still to be able, as they once had, to imagine anything they wanted onto the bottom. The Mid-Atlantic Ridge, for example, was not a heap of detritus the continents left behind when they parted, as Wegener and others had asserted. And the continents were not plowing through the sea floor like icebreakers, with brand-new sea floor forming in their wakes. The Ridge was emphatically the youngest part of the ocean, and still active. Besides, Ewing remarked, characteristically, "They can't move their continents through our ocean floor without leaving marks." Bruce Heezen and J. Tuzo Wilson spoke at different times of an expanding earth, in which the growing Mid-Ocean Ridge and the separating continents were the results of the growth of the earth's surface as a whole. Ewing, Sir Edward Bullard, and Harry Hess thought thermal convection cells in the earth's mantle were making the ocean widen at the Mid-Ocean Ridge, specifically the rift valley, while an equal amount of crustal shortening and destruction occurred in mountain belts and trenches, according to Hess. Henry Menard preferred a similar but more limited process that did not displace continents. In their first paper describing a worldwide Mid-Ocean Ridge, Ewing and Heezen had written: "It must be borne in mind that the rift zone may be the primary feature of the combination, and the ridge simply a consequence of the rift." In a later paper, Heezen and Ewing said the history of the rift's development was contained within its length. The first stage is the African Rift Valleys, which are ten to fifty miles wide, the next the Gulf of Sinai. Then the Red Sea. Later the Norwegian Sea. Finally the Atlantic Ocean. "In the Jurassic, the Atlantic looked like the Red Sea," Bruce Heezen once remarked. "It was just a ridge in a sea. By the Cretaceous, it looked like the Gulf of Aden, wider, the ridge getting buried in sediment. At the end of the Cretaceous, seventy-five million years ago, it was as wide as the Norwegian Sea." While some geologists believed the Mid-

Atlantic Ridge and the other familiar sections of the Mid-Ocean Ridge were just another mountain range, like the Andes (indeed, some geologists did not yet believe there was an oceanic crust thinner than that of the continents, though the government had put up the money to drill a Mohole through it), there was plenty of seismic evidence to the contrary, made on cruises of *Vema*. John and Maurice Ewing wrote a paper about the composition of the Ridge. "Johnny and I were able to say the Ridge is not at all like mountains on the continents — which are belts of folded sediment. Beneath a thin layer of basalt, the Ridge is composed of rock about halfway between basalt and mantle rock, something found nowhere else." The evidence was unpalatable to many. Marshaling all the evidence, Heezen and Ewing wrote another paper that lacked their earlier diffidence. "We wrote that paper just to get everyone off our backs," says Heezen, "and it did. They just forgot about us."

After the oceanographic congress, Harry Hess and Robert S. Dietz wrote papers describing a variation of continental drift which was in accord with everyone's new information about the sea floor. It was a daring variation, when moving continents were considered improbable, because instead of modifying Wegener, they upped his ante. Hess called his an "essay in geopoetry," suggesting a good deal of uncertainty; Dietz named the process "the spreading sea floor." Hess and Dietz's continents, unlike Wegener's, moved slowly along with the entire sea floor, instead of through it, and only as a side effect of the sea floor's own motion. Convection in the mantle moved the crust about. The sea floor spreads away from both sides of the Mid-Ocean Ridge, where new crust is formed from the rising magma of the convection current; the convection currents descend at trenches and mountain ranges, and the crust riding atop them is crumpled together, making room for what is

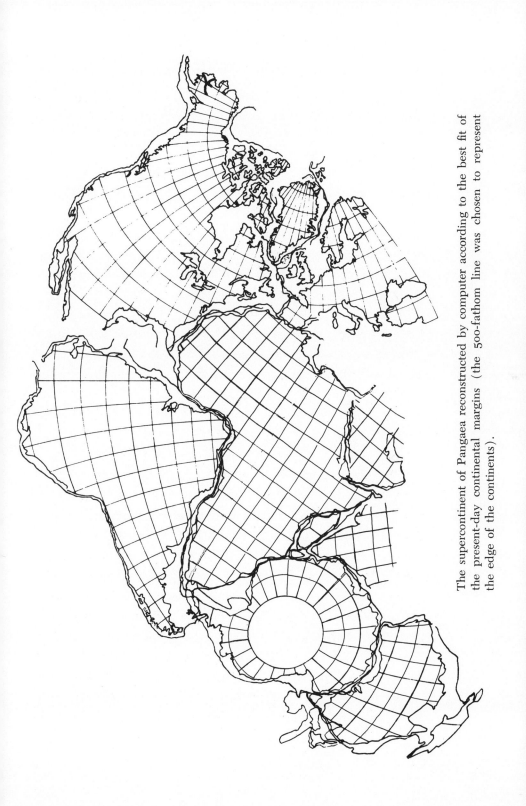

The supercontinent of Pangaea reconstructed by computer according to the best fit of the present-day continental margins (the 5oo-fathom line was chosen to represent the edge of the continents).

manufactured at the Ridge. This mutation was generally accepted as the form in which continental drift was to be proved or disproved. Ironically, the Mid-Ocean Ridge itself was one of the chief causes for skepticism. It was easy to see how the Atlantic could widen. But it was difficult, with a worldwide ridge, to see how there could be widening everywhere, east-west here, north-south there, other directions in between. Africa and Antarctica are encircled by the Ridge; with no intervening trenches or mountain ranges, they should be getting smaller. The North Pacific is surrounded by the encroaching continents of Asia, North America and Australia; presumably it is an older ocean than the widening Atlantic, but the sediments of the Atlantic are deeper. The trenches are where the crust supposedly was dragged under, like ducks by snapping turtles, and in the fifties Hugo Benioff at Caltech mapped zones of earthquake activity beneath the trenches — now called Benioff zones — which the advocates of the snapping turtle claimed proved their point. But Lamont also published evidence, from gravity surveys in the fifties, which seemed to show that the crust was pulling apart at the trenches, not coming together. To solve such dilemmas Bruce Heezen, who with Ewing and others signed some of the gravity papers, has championed the idea that the earth is expanding. He maintained in the fifties, and has continued to maintain, that while the growth of the Mid-Ocean Ridge is fairly straightforward and well documented, it is not possible to find compensating reduction of the crust on an equal scale — several cubic miles a year. Others have picked up the idea from time to time, and there are venerable quantum equations which permit an expanding earth. The idea's fatal flaw is requiring expansion more or less constantly since the earth's formation, while the Mid-Ocean Ridge and the ocean basins are demonstrably very young. In the sixties, however, there was room for considerable uncertainty. Indeed, papers suggesting global expansion continue to be published.

"There was a kind of marvelous tension for several years," says a scientist at Lamont, "knowing the ocean must be very young, knowing the Ridge was there, cooking. You had a feeling every time you went to sea that there were secrets there and we were very close to the answers, and that this cruise might sail right over the answer." It was so simple to say magma rises out of the earth at the Rift Valley and the sea floor spreads away from the Ridge — carrying the continents along — and later disappears into the trenches. But the ocean floor did not look simple, even — perhaps especially — the Ridge itself. The closer marine geologists looked, the more complicated the entire ocean floor began to appear; and the simple pattern of convection and sea-floor spreading began in a few years to look hopelessly naïve. For six years *Vema* and *Conrad*, Lamont's ships, prowled the oceans. They began crossing the Indian and Pacific oceans on world voyages, and Lamont began to write with as much familiarity about the Coral Sea as the Caribbean. During her first round-the-world cruise, in 1960, *Vema* found the Mid-Ocean Ridge on a dozen crossings in the Indian Ocean off Africa and below Australia, remote areas scarcely touched by research ships — or by any ships at all. By not concentrating his voyages on specific areas as other people did, Ewing found he could get a ship to any area within a year of its becoming interesting, while he was acquiring a comprehensive view of the world and information about areas before they became interesting.

In 1960 Ewing's brother John developed the seismic reflection profiler with which Lamont began to see for the first time the disposition of ancient layers within the ocean-bottom sediments — ocean floors of past ages — and the shape of the crustal basement beneath all the sediments. The layers could be followed and mapped with the profiler just as the present sea floor is followed and mapped with echo sounders. With enough profiler profiles, it could be possible to tell what the sea floor was like when the layers, and the basement, were formed — how wide the ocean was, for example — and some of what has happened to it since. To a man who had spent his life trying to piece together what the ocean floor is like, the chance to see what it was like right back to its beginning was unbearably exciting. Profiler records are more fascinating than any other kind of geological record. Even the land-bound geologist never sees much of his subject so plain. "I told Johnny when I saw those first records, 'We can survey the oceans with this,'" Ewing said later. He devoted himself to getting maximum data, and in a few years he had a hundred thousand miles of records — at cost of much effort. He never believed in waiting around for better methods. Every two minutes — instead of every half hour — the watch on Lamont ships lighted a half pound of TNT tied to a balloon, and hurled it over the stern. The string of hydrophones trailing behind the ship was cast loose for an instant to lie still while the echoes came from the bottom. The ship steamed on at ten knots, a train of geysers behind her. Before the next shot the hydrophones were hauled in with a winch; the routine was more than winch motors, at least, could bear, for they regularly wore out. (During the sixties the hydrophones were so improved that they could be towed normally, and, in lieu of TNT, a pneumatic air gun was developed which fires twice a minute instead of every two.) From Cape Horn Ewing tried, in 1962, to follow one of the layers in the sediment as far as the Mid-Atlantic Ridge, but he was repeatedly beaten back by storms. (When he returned home, Lamont's seismographs, like Indians with their ears to the ground, had heard the waves from those storms pounding down upon the beaches of West

Africa.) The layers in the sediments have now been traced across vast areas of both the Atlantic and Pacific oceans, where they appear to be continuous. Two conspicuous ones look like buried abyssal plains, completely flat and of enormous extent, suggesting some oceanwide or worldwide geologic events that resulted in turbidity currents everywhere for age after age. But the sea itself must have been pretty quiet back then, Ewing says, for above the uppermost layer are dunes, mounds, hills and ridges of sediment laid by ocean currents, as well as the abyssal plains of today. In the North American Basin, the Mid-Atlantic Ridge vanishes under the abyssal plains some six hundred miles east of Bermuda. The profiler shows it buried beneath the sediments, and, successively, farther to the west, also under Layer A, an abyssal plain of forty million years ago, and Layer Beta, a plain of the Cretaceous or Dover Chalk age, eighty million years ago. But the profiler cannot penetrate the lowest layer, B, which is itself beneath a mile of sediment. In effect, the profiler shows that if there were no sediment hiding things we would call the entire ocean floor the Mid-Ocean Ridge. To Ewing, the flatness of the old layers was a strong argument that they had not been disturbed since they were formed — and how could they have been shoved hundreds of miles by sea-floor spreading without being in the least rumpled? (It was a reasonable idea — it's hard to push a long noodle, one geophysicist has said — but it was misleading.) This remains one of the mysteries, or miracles, of sea-floor spreading. The profiler also shows that there are even more buried layers just as flat — they look like puff pastries — at the bottom of trenches; and here, according to the drifters, an irresistible force was meeting an immovable object. The profiler even found trenches that had been completely filled by sediment, on both sides of Panama, and south of Puerto Rico, which has an open trench on the north. Such findings did not win new supporters for sea-floor spreading. Ewing and Lamont stalked and snared the lovers of sea-floor spreading for nearly five years. They had the data. Indeed, they had too much.

The Ridge, too, became more complex. In 1962 Lamont and Woods Hole ships found that the Mid-Atlantic Ridge is offset as much as several hundred miles along fracture zones, or faults, and they soon located six of these offsets. Huge fracture zones in the Pacific had been mapped by H. W. Menard for almost ten years, scarps several thousand feet high and three thousand miles long, apparently displacing the ocean floor many hundreds of miles; but they had not been associated with the Mid-Ocean Ridge. In the Atlantic, American research ships habitually had sailed east to west, parallel to the fracture zones, and had never noticed them. Russian ships, however, were forbidden United States ports and had to sail up and down the ocean between Europe and hospitable South American harbors; they noticed the Atlantic fracture zones first. That the Ridge is offset, Heezen and Ewing wrote in 1962, "is a fact of certain but unknown significance." It soon was clear that the Ridge does not wind its way through the oceans sinuously like a snake but zigzags like stairs; between the fracture zones that offset it sideways the Ridge is straight. It also became clear that, as with a stairway, the direction of the whole was not always the direction of the parts — and that the orientation, and even the location, of the Ridge in most of the world was even less well known than had been supposed. There was confusion, for a while, in many geological camps. What if the Ridge were so oriented between Australia and Antarctica, say, that it could never have separated those two continents? (As a matter of fact, in the Indian Ocean where the Ridge points north, its segments point northwest, and the fracture zones point northeast.) With vast lengths of the Ridge virtually unseen, and offset by unknown numbers of fractures, exploration concentrated briefly, as it had several times before, on the study of earthquakes. And by coincidence there already existed new computers and a new worldwide network of seismograph stations that could locate earthquakes precisely enough to reveal the steps and offsets in the Ridge (yet earthquakes are too spasmodic, both in time and place, to reveal more than a tantalyzing sample of the whole picture).

The gravitational and magnetic fields of the earth have only recently been extensively used to study it, though interesting things like ore deposits have been discovered through local variations that they cause in the fields. "We've been the main pusher of gravity and magnetics at sea," said Ewing. He began towing magnetometers behind ships in 1948, getting continuous measurements of the field when the machine consented to work. Since then every Lamont cruise has towed one, and work that began out of no more than a sense of duty and obligation to use every tool available became painfully exciting. In 1952 R. G. Mason, of Scripps, borrowed Lamont's magnetometer and towed it in the Pacific, later building his own. Lamont found an obvious magnetic high over the rift valley, but — that apart — for years it was unclear what the local variations were that the magnetometer was responding to. Enthusiasm waned so at Scripps that the opportunity of letting a government ship tow the magnetometer was nearly ignored. The ship was making a detailed survey — steaming courses five miles apart — of a two-hundred-by-fourteen-hundred-mile area between San Diego and British Columbia. When Mason plotted the first magnetic data, he had a startling pattern unlike anything seen before, stripes of high magnetic intensity alternating with stripes of low intensity. As the ship continued in its monotonous way, the pattern filled the entire quarter million square miles it surveyed. Three fracture zones cut across the stripes, and the stripes were different on each side of them. Across the southernmost fracture, the Murray Fault, a match was found, the stripes to the south being eighty kilometers west of identical ones to the north. A ship was sent out along a second fault in 1958, and a match of magnetic stripes found after a hundred and thirty-five kilometers. "This obviously indicates," three scientists wrote in 1963, "that after the formation of the geomagnetic anomaly pattern, a fault movement along this line caused an east-west displacement of 135 kilometers." Several hundred miles of survey along the remaining fault the following year failed to disclose a match. In 1960, a ship went out farther still, and the match was found, displaced six hundred and twenty kilometers. Someone re-marked that if displacements that large could occur, continental drift was not all that fantastic. But, if these were displace-ments, they were odd. Two of the faults were apparently displaced in one direction at one end and in the opposite di-rection at the other end. It was reported that while the pattern north of the Mendocino fracture (way out to sea) was, relative to the south, six hundred and twenty kilometers to the left, near the California coast it was fifty-five kilometers to the right. If the crust was displaced, it also had undergone some other contortions, of which there was no sign. Moreover, out to sea the fractures merely petered out into ordinary, unfractured ocean crust, presenting a topological problem, as though all Manhattan west of Fifth Avenue lay fifteen blocks north of the East Side, West Forty-second Street running straight into East Fifty-seventh Street without any change in the shape or shore-line of the island. If the faults appeared to defy rational expla-nation, the striped magnetic pattern stimulated the purest speculation. A similarity between the pattern and strain pat-terns observed in plastics explained nothing, but was commented on, approvingly, by advocates of continental drift — though they did not specifically associate the pattern or the fracture zones with sea-floor spreading. There appeared to be no physical connection with the Mid-Ocean Ridge, and no one apparently tried very hard to assert one. Alternating strips of magnetized crustal rock were proposed to explain the anomaly pattern, but were themselves in need of even more explanation. "If they hadn't been messed up with the continent," says Ewing of the magnetic anomalies off California, "they would have been understood immediately."

The magnetism of the earth has been described but not explained; experts complain it is the planet's most described and least understood property. William Gilbert, physician to the first Queen Elizabeth, showed that the earth's field behaves like that of a simple bar magnet, or dipole. He carved a magnet into a sphere, observed the field on its surface with iron filings, and found it similar to the earth's. This is still as good a description as any. For several centuries, it was assumed that the earth is a permanent magnet, but it is not. Rocks and metals that are permanent magnets lose their magnetism when heated beyond a certain temperature, called a Curie point. The earth is both hot and liquid inside, the embarrassing discovery of the turn of the century that left the field apparently existing for no reason. The best bet today is that the field is produced by some sort of fluid dynamics in the core — which the magnetohydrodynamicists will hit upon eventually. Mars has no core, and no field; neither has the moon. When a lava cools below its Curie point the ferrous particles in it align themselves with any magnetic field present — the earth's in the absence of anything else — and thereafter are a sort of fossil of that field at that time. Measuring a fossil field is an exceedingly delicate operation that was not often performed before the nineteen fifties; but when done it revealed there have been significant changes in the field. An English group found that the relation of the North Pole to India has progressively changed by twenty-five hundred miles since the Jurassic period a hundred and fifty million years ago, either because the pole moved or India moved, or both (or because, according to Sir Harold Jeffreys, the rock samples were affected by being broken off with a hammer). By 1955, a group of researchers at Newcastle developed, with a huge sample of lavas from all of Europe over a span of hundreds of millions of years, a consistent magnetic history of that continent which was hard to argue against. S. K. Runcorn, the leader of the group, wrote: "Appreciable polar wandering seems indicated. There does not as yet seem to be a need to invoke appreciable amounts of continental drift to explain the paleomagnetic results so far obtained." Two years later, however, Runcorn had a path of polar wandering for North America as well. Its shape closely resembled the European path. But from the pre-Cambrian to mid-Mesozoic the paths were consistently thirty degrees apart; thereafter they converged, and for the last few tens of millions of years they were superimposed. The tale was the same everywhere within each continent, differing only between them. Perhaps, for most of the time that life was evolving, each continent had its own private magnetic pole, and they merged their resources only recently; but perhaps, instead of two poles merging, the continents separated. Wegener had expected such findings, but his paleomagnetic evidence was suggestive rather than conclusive; the new evidence was a great deal better and had begun to revive talk about continental drift when the Mid-Ocean Ridge and its rift were discovered.

Even more bizarre than a planet with wandering poles is a planet with reversing poles. Seven hundred thousand years ago, and for nearly a million years before that, the north magnetic pole was in Antarctica; and it will eventually return there, creating navigational problems. In accordance with the general mystery of the magnetic field, why it makes the excursions it does is thus far better known to itself than to magneticians, though it leaves plentiful signs of its passage. Reversely magnetized lavas were discovered in India in 1855 ("In retrospect, the most surprising thing about [the] discovery of reversely magnetized rocks," begins an article in *Nature*, "was that it should have evoked so little surprise"), in France in 1906, and in Japan in 1929. In the nineteen fifties it became clear that at least a large minority of rocks were reversely magnetized. Rather than suppose that the earth's field might reverse itself, researchers suggested that the rocks (like fossils) were *sui generis*, self-reversing; several theoretical models were offered, and several self-reversing rocks discovered and induced to reverse. However, in the early sixties three workers, Allen Cox, Richard Doell, and Brent Dalrymple, of the United States Geological Survey, reported on large numbers of sample rocks from places all over the world that embodied the same sequence of nine reversals at the same times (from radioactive dating) during the last three million years. At one time, all fresh lava flows had been reversely magnetized everywhere, at another time all normally magnetized. Sediments heated by the lava had the same polarity as the lava. Today, most scientists agree that the field reverses from time to time, though an occasional rock may perform the service for itself. But as Cox and his associates wrote: "The idea that the earth's magnetic field reverses at first seems so preposterous that one immediately suspects a violation of some basic law of physics, and most investigators working on reversals have sometimes wondered if the reversals are really compatible with the physical theory of magnetism." Neil Opdyke, a paleomagnetician at Lamont, recalls that in the sixties his kind often endured being

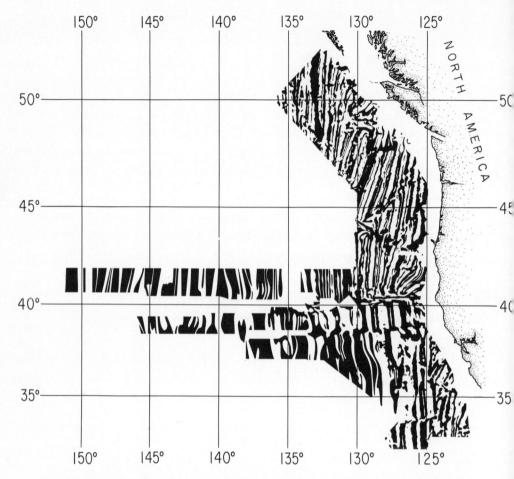

The pattern of magnetic anomalies found off the West Coast. Positive anomalies are black, negative anomalies white. The Mendocino fault lies near 40 degrees; the Pioneer fault near 38 degrees; and the Murray fault near 34 degrees. A negative anomaly is offset by the Murray fault from 127 degrees to 124 degrees.

NORMAL

REVERSED

NORMAL

REVERSED

Changes in the earth's magnetic field are recorded in the rocks.

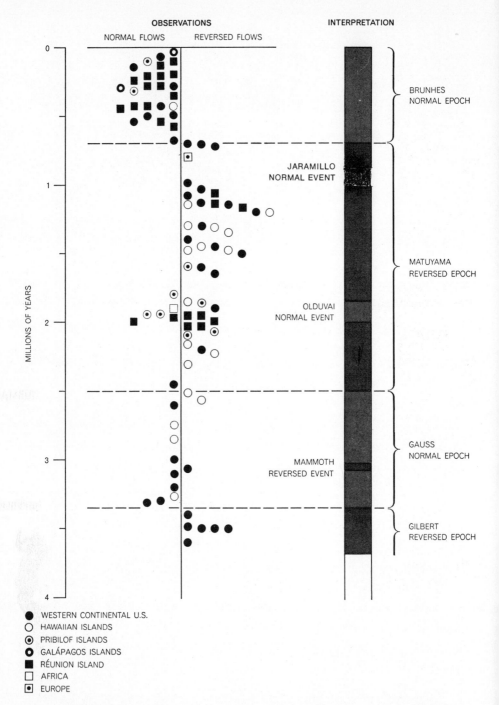

A chronology of how the earth's magnetic field has reversed itself. Volcanic rock samples from nearly one hundred locations in both hemispheres fell at first into six groups — four long epochs and two short events. The Jaraomillo event was discovered later.

called "paleomagicians." But the physical theory of magnetism does in fact allow reversals. By drilling through accumulated lava beds one gets a vertical history of reversals. Chronologies have been constructed and are still being extended; detail keeps being added. The poles have switched position with some regularity for nearly a thousand million years, the present extent of the evidence. The field was almost continuously the same for some fifty million years about the time the first vertebrate fish were appearing, and again when dinosaurs were disappearing. Recently, however, reversals have been relatively frequent. There appear to have been cycles of reversal frequency (and also of predominantly reversed or normal polarity) lasting several hundred million years. Although the present state of affairs is understandably called normal, reversed periods seem to have accounted for a greater part of the world's history. By and large lately, the field has been staying the same for about a million years, a period paleomagneticians have named an epoch; but the epochs may be punctuated by shorter periods of opposite polarity called events, which last only tens of thousands of years. The present period is an epoch that has lasted nearly seven hundred thousand years and is called the Brunhes, after an early finder of reversely magnetized rocks. The preceding epoch, the Matuyama reversed, lasted a million and a half; before were the Gauss normal and Gilbert reversed, each lasting about a million years. In the midsixties, when the known history of reversals extended back only about three million years, just two events had been discovered — the Olduvai normal event, near the beginning of the Matuyama epoch, and the Mammoth reversed event, in the Gauss epoch.

In the fifties and early sixties, attempts at Lamont to show polar wandering or continental drift with samples from seamounts proved inconclusive. But research ships each year

produced tens of thousands of miles of magnetic profile of the
ocean floor, and there clearly was a distinctive pattern across
the Mid-Ocean Ridge in the North and South Atlantic and
Indian oceans. What the pattern was in aid of, however, was
not clear. It reflected topography, but it reflected more. A
seamount produced a strong positive anomaly, unless it produced a strong negative anomaly. There was a whopping
anomaly at the crest of the Ridge, whence the ups and downs
gradually subsided toward the flanks. In 1963, a few months
after Cox et al. published some of their evidence for field
reversals (but before there was a clear calendar of them),
F. J. Vine and D. H. Matthews, of Cambridge University, proposed that the magnetic anomalies of the Mid-Ocean Ridge
were caused by a combination of reversals of the earth's field
and sea-floor spreading. They wrote in *Nature*: "If spreading
of the ocean floor occurs, blocks of alternately normal and
reversely magnetized material would drift away from the centre of the ridge and parallel to the crest of it." Dietz, too, had
described "strips of juvenile sea-floor" forming at the Ridge;
and these, differently magnetized, would create a magnetic
pattern like the magnetic stripes found in the east Pacific off
California. The sea floor would record reversals of the field as
tree rings record wet years and dry years. This was a remarkable piece of imaginative thinking, but no one seemed to believe it might be true. Vine and Matthews had some magnetic
profiles of the Ridge in the Indian Ocean, and by program-
ming a computer with their theory they produced theoretical
profiles that were fairly similar to the real ones. The demonstration was considered interesting, but not good enough to be
convincing. In the meantime, a United States Navy aeromag-
netic survey, Project Magnet, made fifty crossings of the Reykjanes Ridge, the segment of the Mid-Ocean Ridge immediately
south of Iceland, and Lamont obtained access to the data. A

corollary of Vine and Matthews's idea was that each side of the Ridge would be the mirror image, magnetically, of the other. When the Navy crossings were correlated they produced a pattern of parallel stripes, like those in the east Pacific; and one side of the Ridge was nearly the mirror image of the other, at least in the two-hundred-mile-square survey. The survey, unfortunately, was not improved by the military mind; the plane crossed interesting territory between Iceland and the survey area with the magnetometer turned off. But no one had ever correlated magnetic profiles of the Ridge before; if the east Pacific pattern was a characteristic of the Ridge it was a major discovery. Ewing got *Vema* to the Reykjanes Ridge in very much less than a year. "I told Manik Talwani, the chief scientist, 'Don't come back until you can tell me if the bands are continuous,'" Ewing said. Though the Navy had supplied better evidence for Vine and Matthews than they had themselves, detailed examinations made their idea seem less ravishing. The anomalies were symmetrical about the axis of the Ridge, but the one at the crest of the Ridge — discovered on *Atlantis* in 1948 — was apparently quite a different breed from those farther away, though the anomalies of the Ridge flanks had been created at the crest if sea-floor spreading occurred. No one was able to imagine a mechanism for turning crest anomalies into flank anomalies. There were also certain technical difficulties in producing any anomaly at all by sea-floor spreading. Nonetheless, Vine and Matthews's was the only hypothesis around, and other papers referred to it, if only in such savage (for scientists) terms as "dubious" and "untenable." "Vine and Matthews was interesting, but they hadn't the data to back it up," a seismologist at Lamont said recently. "Not many took it seriously here." Vine himself wrote: "At the time this concept was proposed, there was very little concrete evidence to support it, and in some ways it posed more problems than it solved." Even an ardent advocate of continental drift once dismissed an idea as "the biggest joke since Vine and Matthews."

Each year the Ridge seemed to become more complex, and more puzzling. In 1964, Lynn Sykes, a Lamont seismologist, found that the earthquakes along fracture zones are confined to the part that offsets the crest of the Ridge. To either side, a far greater length of the fault, just as impressive-looking, is aseismic. In 1965, a group from Lamont proposed that the fracture zones and the offsets of the magnetic pattern are not caused by sections of the earth's surface slipping past one another — as first the Scripps researchers, and then everyone else, had assumed — but are characteristic of the Ridge from its beginning. The magnetics, they said, are offset only because the Ridge is offset. Each band is characteristically a certain distance from the Ridge axis. Fracture zones are cliffs because one side is nearer the center of the Ridge than the other, therefore higher — and they are active only where the Ridge is active, the center. Late in 1965, Professor Tuzo Wilson proposed further that the earthquakes are only in the one part of the fracture because the sea floor there, between offset segments of Ridge, is spreading in opposite directions. After examination of the idea of sea-floor spreading, Wilson said, he had found that the fractures are a new class of fault, and that the motion causing the earthquakes would be the opposite of motion that would offset the Ridge. This was a rarity in geology, a prediction that could be checked — by examination of seismograms — though Wilson did not do so.

"It's primarily our data that's led to this whole thing," said Ewing, "and ours that will settle it."

Eleven

Eltanin-*19*

SEA-FLOOR SPREADING was established as a fact and not wishful thinking in 1967 at Lamont. One ideal piece of evidence turned up there — not surprising, considering the hunger with which Lamont gathers evidence — and once the pattern was seen in a single bit it became visible in others, of which great quantities soon were brought out of Lamont's files. For a number of people there was a period of almost unbearable excitement, lasting about two years, and spreading rapidly to other geological establishments. Some people think it the most remarkable period in the history of geology. Today young men speak like old war veterans, with the feeling of having lived through an era the like of which will not be seen again. By the end of 1968, a group of ideas called global tectonics were accepted almost universally among those who had heard of them. By 1970, they had suffused and were reshaping traditional geology; the impetus for mountain-building was accepted as coming from sea-floor spreading, and geologists are now trying to read its history and consequences in the rocks of folded mountains of the west coast of the United States, the east coast, and the Mediterranean. Today, a paper on continental mountain-building and other upheavals that took no note of plate tectonics would be surprising indeed.

One no longer enters Lamont by the steep, switchback driveway to the old Lamont house, or the old narrow road, lined by lawns and hedges, through Sneden's Landing, which has scarcely changed in twenty years. A new tarred road, marked by an empty guard shack and a very small sign, leads off the bumpy country highway through well-manicured woods, past a few small fields and a few clapboard houses to a cluster of low, rambling laboratory buildings, half of them steel prefabs. A fork of the road comes last to the old house of pink sandstone and red clapboards. Before it, in bronze, a youthful Lincoln reads a book astride his grazing horse. There are fruit trees about the grounds, laurels, and magnolias. In the spring Lamont smells of blossoms, in the autumn of windfall apples. The countrified grounds are a relief from the nearby city, but there is no spectacular waterfront, as at Scripps and Woods Hole (Lamont ships dock at an industrial pier a few miles away, rarely more than once a year) and there are few if any sailboats nearby to leave work early for. There is not, of course, even a view of the Palisades. (And when Lamont built a new laboratory with a view of the Hudson, Laurance Rockefeller became unhappy; he subsequently put up $35,000 to paint the face of the building green and plant trees to restore the appearance of the Palisades from across the river.) Lamont has acquired few attractions beyond the excitement of the work, but scientists feel life is intenser there than at other institutions. A visitor has remarked that if an observatory were ever established on another planet it would feel like Lamont, while others have thought of the Cavendish Laboratory, where nuclear physics was born. Many of Lamont's four hundred and fifty people are at sea or abroad. A large number at Palisades is always just back from, or just leaving for, somewhere — the Arctic, the South Sandwich Islands, the manned space flight center, Suva, Singapore.

A branch of the road goes up to the oceanography building on the edge of a bluff commanding the river, which is yearly less visible through Mr. Rockefeller's trees. Before this yellow brick construction roosts *Vema's* old figurehead, a rapacious-looking eagle. Inside, the mystery of the earth's surface began to unravel during the winter of 1965–1966. Lamont then was collectively agreed, with near unanimity, that something of great significance was happening on the ocean floors, and that it was not sea-floor spreading. "Ewing found the oceanic crust, and that seemed to assure the permanence of the continents and oceans," says Xavier LePichon, one of the younger and more prolific men, who has since moved to the Centre National de Recherche Scientifique, in Paris. In press was a series of impressively researched articles, instigated by LePichon, and signed by him and a variety of other authors, on every aspect — magnetism, topography, gravity, temperature gradient — of the Mid-Ocean Ridge; all concluded after painstaking argument of the evidence that it pointed away from sea-floor spreading. "We are not geologists, and misinterpreted the evidence — particularly the seismic profiler — often," says LePichon.

Ewing had been encouraging people to study magnetism, and late in 1965 some graduate students found magnetic-field reversals in sediment cores. For many, it was this discovery that established field reversal as fact — sediments and lavas have almost nothing else in common. "There's a big literature of freak lavas," says Ewing, "but when the cores confirmed, and extended, the reversal story from lavas there was no further doubt, as far as I was concerned. I had tried to get the stuff studied for about ten years. First one man was on it; his conclusions were inconclusive. Then another man. Then Neil Opdyke came; he was doing paleomagnetism, but of old rocks. I gave him a long talk, said you have here an unequaled opportunity; and he played with his old rocks some more."

Neil Opdyke is a solid, careful scientist, who had worked in one of the first laboratories to pursue polar wandering, and he was an early believer in continental drift. Ewing, though he disagreed with all of Opdyke's driftist conclusions, had sought him out and hired him from abroad. Opdyke is enthusiastic about the results his laboratory obtained. "It was one of the high points of my scientific life," he says. "I had asked John Foster, a graduate student of mine, to look at some specimens. He thought he had to get undisturbed cores, and he let the project lag." Foster, a tall, solemn young man, had been deputed to build Lamont a device for determining the magnetic orientation of samples. He had completed his Mark III model of the device — a spinner, as it was called — which looked like an old-fashioned towel roller with some metal grids around one end, and was wondering what to do with it. "One day another graduate, Bill Glass, came in with a friend who was on his way to a new job," says Foster. "His friend was interested in building a spinner like mine. Glass had a piece of mud from a core; his roommate had been studying the fossils in it. Could we detect any magnetism? We tried, and got an unmistakable response. We were excited, because other attempts had failed. We had the luck to have everything work right the first time. Just at closing time we ran to the core lab for more samples. We cut three samples a meter apart from a core that had already been studied and dated, through fossils, from top to bottom. The middle of the three clearly was reversely magnetized. We went to supper and worked into the night. Neil wasn't convinced at first."

"I told them it wasn't enough just to measure," says Opdyke. "You have to demagnetize too. Otherwise, you're open to the charge that you measured a recent addition to the sample, an overlay, the effect of events subsequent to the formation of the sample. The samples had to be magnetically washed of all but permanent magnetism."

"We worked at night until we got good results," says Foster. "Within a week, Neil realized something first-class was happening and took charge. It was the most exciting time I've ever spent. We had an advantage over people studying lavas, because sediments collect continuously, and we filled in gaps in the historic record and found reversals too short to show up in lava flows. We found that the field reverses very quickly, within just a few thousand years. It gets steadily weaker, reaches zero, then increases with the opposite polarity."

"We succeeded in correlating reversals with fauna," says Opdyke. "It was the most important single advance in sedimentology. Through magnetic reversals we can correlate synchronous events in different sediments, with different fossils, all over the world."

"We'd found a universal fossil," says Foster. "So you could tell for sure if an event that happened to clams on this continent occurred the same time as what happened to elephants across the ocean." Among other things, the "universal fossil" has since shown that the ice ages began much earlier than anyone had thought, and that there have been more of them than anyone thought.

During winter, spring, and summer 1966, under Opdyke's direction, the graduate students did all the tedious work of buttressing the findings already made or suspected. They worked through a dozen long cores — sixteen to forty feet each — which had already been painstakingly analyzed, dated, and correlated by fossils. Samples were taken every two inches and measured. When a reversal was found they took continuous samples. One by one, reversals appeared in each core as they worked down, were dated by fossils, and found to correlate not only core to core but to the existing chronology established from lavas. The cores reached much farther into the past than did the chronology from lavas, and Opdyke and his group slowly approached a time when they would leave familiar ground.

Before that happened, there was excitement across the hall, where the magnetic pattern of the sea floor at last began to make sense; and the two investigations began to influence each other. In January, Walter Pitman, a graduate student (who has since become one of Lamont's senior scientists) with an office opposite Opdyke's rooms, got three magnetic profiles of the Mid-Ocean Ridge from Lamont's computer, which had reduced them from a great many yards of paper to convenient size and had subtracted the normal effects of the earth's field, leaving only the interesting aberrations. He returned to his office and began to look them over. They had been made in the far South Pacific aboard the National Science Foundation's ship *Eltanin*, which (until she was laid up in 1973) hovered about Antarctica and did marine geophysics under Lamont's direction. Pitman had only recently returned from a voyage on her, during which two of the profiles were made. Ships, even research ships, rarely visit that remote part of the ocean, and information about the Mid-Ocean Ridge there was scarce. Great pains had been taken while the profiles were made that the ship's course should be straight, and perpendicular to the Ridge, in order not to distort perspective. Nonetheless, Pitman approached the profiles without great anticipation; there was no reason to expect anything very unusual. He and his boss in the marine magnetics department, James Heirtzler (who since has moved to Woods Hole), had already sent off for publication the abstract of a paper they planned to deliver at the annual meeting of the American Geophysical Union in April. It only promised three well-made new profiles from the South Pacific. It looked so routine (Pitman says) that hardly anyone came to hear the paper he did give — which was quite something else — and most of them had been tipped off by Neil Opdyke. Pitman glanced over the two profiles made while he was on *Eltanin*, and passed on to the third, made on the previous voyage, *Eltanin*-19 (by which name the profile has come to be known, though some of Pitman's colleagues call it the Pitman profile). That one stunned him. Even to an untrained eye, the *Eltanin*-19 profile does not look haphazard or coincidental. To one who for years has tried unsuccessfully to stare meaning out of data, it has a special excitement. It is too smooth, too even, the variations of one side too faithfully reproduced on the other. Even to eyes glazed by the brassy interchangeabilities of mass production, its perfection is visible. Later on, when it was shown about, a man at Lamont said, "This is so perfect that I know sea-floor spreading couldn't have done it." Norms are human obsessions; nature rarely bothers to construct one. The profile is probably the most remarkable piece of evidence to come along in any field in some years. It consists of vigorous peaks and valleys (of magnetic intensity) arranged in obvious symmetry, size for size, cluster for cluster, about the center of the Ridge. To an educated eye it embodied, almost too well, the Vine and Matthews hypothesis and sea-floor spreading.

"It hit me like a hammer," Pitman said recently. He is an

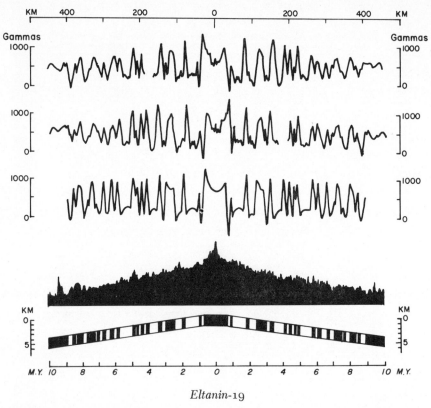

Eltanin-19

ABOVE. *First:* The mirror image of the magnetic profile made over the Mid-Ocean Ridge during *Eltanin* voyage 19. *Second:* The profile itself. *Third:* A profile generated by a computer programmed with the assumptions of sea-floor spreading. The topography of the ridge is profiled in black above the model of field reversals and sea-floor spreading which was given the computer.

BELOW. Sketch of the Ridge showing normally and reversely magnetized sea floor.

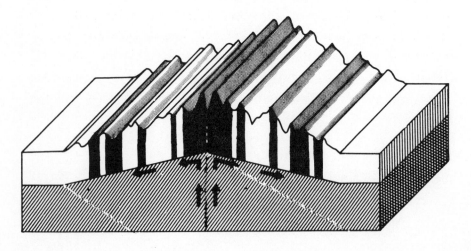

affable man in his late thirties, balding, and with luxuriant, drooping orange mustaches. His office on the second floor of the oceanography building looks over a steep slope and the Hudson. There is just room to sit between his desk at the window and the bookcases against the other walls; papers and scientific periodicals are piled everywhere. "In retrospect, we were lucky to strike a place where there are no hindrances to sea-floor spreading. There's no other place we get profiles quite that perfect. There were no irregularities to distract or deceive us. That was good, because people had been shot down an awful lot over sea-floor spreading. The symmetry was extraordinary. I had thought Vine and Matthews was a fairly dubious hypothesis at the time, and Fred Vine has told me he was not wholly convinced of his own theory until he saw *Eltanin-*19. It does grab you. It looks just the way a profile ought to and never does. On the other hand, when another man here saw it his remark was, 'Next thing, you'll be proving Vine and Matthews.' Actually, it was his remark that made me go back and read Vine and Matthews. We began to examine *Eltanin*-19 and we realized it looked very like a profile that Vine and Wilson published just before our data came out of the computer. Their profile was from the North Pacific, from the Juan de Fuca Ridge, a segment of the Mid-Ocean Ridge which Wilson had just identified that year off the Juan de Fuca Straits south of Vancouver Island. We said, 'Isn't this interesting.' If the pattern was the same eight thousand miles away, it was the more reason for thinking it was determined by something worldwide like field reversals — Vine and Matthews's idea. Vine and Wilson had tried to relate their magnetic profile to the known history of magnetic reversals, but they failed — in part because, as it turned out, that history still was incomplete. One event, the Jaramillo, hadn't been found yet. Without that, you couldn't get a fit without false assumptions or distortions.

But we were lucky. Just then, Opdyke began to put together the history of reversals from sediments, and pretty soon he found the Jaramillo." (Ruefully, Vine wrote, later on: "It will be seen that, had the authors had more faith in the idea . . . they could have predicted the Jaramillo event.")

Near the end of February, Opdyke and his graduate students found an undescribed magnetic event in some North Pacific cores. "We called it the Emperor event after the North Pacific group of seamounts," says Opdyke. "Then Vine came through one day and said, 'I see you've found the Jaramillo event.' Damn! There goes another breakthrough! Someone else just found it too. Vine and Wilson thought the Jaramillo was the Olduvai event, which is the next youngest event but almost twice as old. That threw off their entire pattern. With the extra blip, the history of reversals explained the sea-floor pattern easily. Either that, or God was playing tricks on us."

"Now we had a good time scale," says Pitman, "and we used it to make a model of recent sea-floor spreading in the South Pacific. It was the crudest kind of model — just alternate blocks of normally and reversely magnetized sea floor. We decided arbitrarily that the magnetized layer is two miles thick, whereas it now appears to be less than one. We also assumed a spreading rate of four and a half centimeters a year on each side of the Ridge. It was crude, but it fit. We programmed a computer to produce a profile from it, and the profile almost exactly matched *Eltanin*-19. That gave us confidence to make a profile for one centimeter a year, at different latitude and longitude, and that matched the magnetic profiles of the Reykjanes Ridge, below Iceland. Lamont published those with arguments why they didn't fit the Vine and Matthews hypothesis, but everyone had remarked on their rough symmetry. I don't think Vine himself had the nerve to claim the Reykjanes Ridge then. There's no rift valley there — or in the South Pacific —

and everyone thought rifting had to go on in a rift valley. It's one of the ironies that the rift, which started everyone thinking about continental drift again, isn't found where sea-floor spreading is working most smoothly — just where it's slow and difficult. Slow spreading makes for the big topography, but the beautiful, even pattern comes only with fast spreading — and puny topography.

"After the Jaramillo was found, we knew what a profile meant, and what we could do with it. The magnetic history that had been worked out in lavas only went back a few million years and occupied only a very small central part of our profiles. Heirtzler calculated that the anomalies farthest from the center must be about eighty million years old, if the sea floor had always been spreading at the same speed. And that meant that everybody's ideas about how old the ocean floor must be were completely wrong, because in the South Pacific at least we were talking about almost all the ocean floor there is. We had to translate the remaining anomalies into eras of normal and reversed magnetism and show the same thing worked in every ocean all over the world. If they did, good; but if not, the whole theory might be wrong and we were deceiving ourselves. We weren't really thinking at all about what it meant, until we were through — except for Heirtzler's immediate conclusion that the ocean is practically all brand-new. We were getting out all the magnetic profiles that had ever been made across the Ridge, and there was a bunch of us looking at them. For a few months Neil's group and ours practically lived in each other's laps. His pattern, that he was getting from sediments, and our pattern were always the same. By spring, he'd gotten back to the beginning of the Gilbert reversed epoch, three and a half million years. That was as far as Cox and his group had gotten with lavas. But just a little farther on in our profiles there was a conspicuous group of four high, narrow anomalies that I had called Four-Fingers Brown after a famous gangster in Newark, near where I grew up. One afternoon in June, Neil and I together ran the samples from an entire core through the spinner, and at the bottom of the core the reversals came booming in, Four-Fingers Brown. That was the first confirmation we had of our sequence. Until then we'd been showing that our pattern matched someone else's. This time, it was the other way around." The underpinnings of any science are made of shared experiences, whose discovery makes science exciting to scientists. The same tale, embodied in the age of lava flows, and vertically in deposits of mud and clay, and horizontally on the flanks of the Mid-Ocean Ridge, is a level of identity never achieved in geology before; it gave geologists a great deal to do, and gave their statements a new assurance.

In June of 1966 — after Four-Fingers Brown — Pitman and Heirtzler were sufficiently satisfied with their findings and provings to call in a few people to look at what they had. Two, Jack Oliver and Lynn Sykes, were seismologists and taught in Columbia's geology department. "Sykes and Oliver," says Pitman, "went roaring back to look at their data."

"It was beautiful," says Sykes, a quick young man with heavy eyebrows and a professorial tonsure. "They'd made transparencies of a thousand-mile length of the *Eltanin*-19 profile. By taking two of them, turning one around and putting it over the other, you saw vividly the symmetry of the profile. Some other people's work had been suggestive, but it was the Pitman profile that really made people believe Vine and Matthews, which hit you that this really was so. I don't think they really quite realized the full import of what they had when they called several of us in. At the time I had been trying to think of ways to test Wilson's idea of a new kind of — transform, as he called them — faults, but hadn't got started. I had

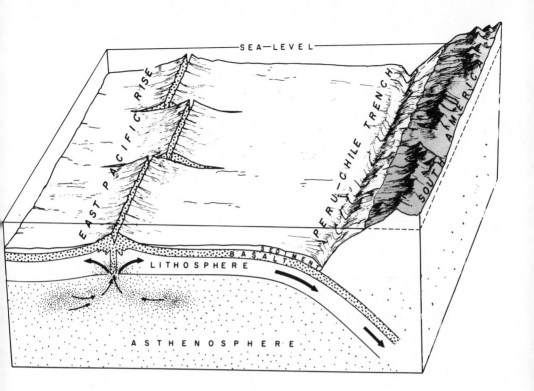

Sketch of the motions of sea-floor spreading. Material from the mantle rises at the rift valley, where the sea floor is separating. Sea floor is consumed at the trenches, mostly returning to the mantle, but in part also melting and rising into the mountains above. The Ridge is offset by transform faults. It was thought at first that the entire Ridge and sea floor had moved at the faults, creating offsets, for example thus:

That this is not the case has been a compelling reason for believing in sea-floor spreading. In fact, only between offset portions of the rift valley do the Ridge segments slide past one another, creating frequent earthquakes, and in the opposite of the expected direction:

Beyond the offset is a fracture zone, but seismic activity subsides (the arrows are in unison).

been in Fiji. When I came back, during the winter, someone told me about Wilson's theory. I saw the profile in June. The next morning I went to work. Essentially I knew what I had inside a week. I looked at twenty earthquakes and they all worked. They had no right to. It's a matter of whether the first pulse on the seismograph is up or down, something only the most recent instruments have been able to distinguish. It shows you the direction of motion involved in the earthquake; and in transform faults that is exactly the opposite of what classical geology predicts for horizontal fault motion. The old idea was that if something is offset to the left then it has moved to the left, and the motion on the intervening fault will be to the left. But if it's the Mid-Ocean Ridge that's offset, then the sea floor spreads away from the crest, back along the fault — to the right. And that is what the first-motion studies of earthquakes show — right lateral motion where a classical interpretation predicts left lateral. I was a little hesitant about presenting my results. It was a matter of trying them out on people first; and the more I tried, the better things sounded. Sea-floor spreading was the only thing that could produce those earthquakes."

While Lynn Sykes was assembling his confirmation of the first prediction made from sea-floor spreading, he, Jack Oliver, and their colleague Bryan Isacks were able to show with earthquakes that one large segment of the earth's crust is being pushed, or drawn, under another at the submarine trenches, and also at certain mountain systems like the Alps and Andes. "In the early sixties," says Sykes, "seismology here was on a vein — surface waves — that was about mined out, and we had decided, Oliver, Isacks and I, to set up a study of Tonga, because it has so many deep earthquakes — though it was not clear then what problem this would solve. Deep earthquakes were first observed in the nineteen twenties, but they weren't believed in until the thirties. To get an earthquake, you have to have something hard that can rupture. Down deep, in the mantle, it was thought, high temperature and pressure should melt everything. The Japanese showed that, while deep earthquakes might not be theoretically possible, they occur. In 1964, when we set out our first instruments in Tonga, it was still a mystery why. They only occur under trenches and island arcs. They lie in distinct zones dipping down from the trenches, under the island arcs — which are volcanoes — to a depth of six hundred miles. In our first papers we thought the zones were about a hundred miles thick, but we found they really are only about fifteen miles, very thin indeed. After we put in our instruments around Tonga, we noticed right away a few things, like — this was Bryan Isacks — the high frequencies were traveling along the plane of the zone. This we didn't expect at all. High efficiency in transmitting waves means hardness, coldness, all that the mantle was supposed not to be. So we knew we had to revise our ideas about what the zone is like. It all came together in a model late in 1966. By then all our thinking was influenced by what was going on in magnetics over sea-floor spreading. Jack Oliver had been toying with the idea that this zone was a piece of something different hanging down there. Well, waves also propagated very easily along the sea floor fifteen hundred miles to our Raratonga station. The sea floor is hard and cold. All of a sudden, it became very simple and obvious. We were led to the conclusion that the earthquake zone is the same as the sea floor, and then — Pitman had just given evidence for sea-floor spreading — that the sea floor was being pushed, or pulled, down into the mantle to form the zone." The pushing or pulling was causing the earthquakes. With less than usual scientific diffidence, Isacks and Oliver wrote: "The evidence is largely qualitative, but it is not subtle and it is definitive. The

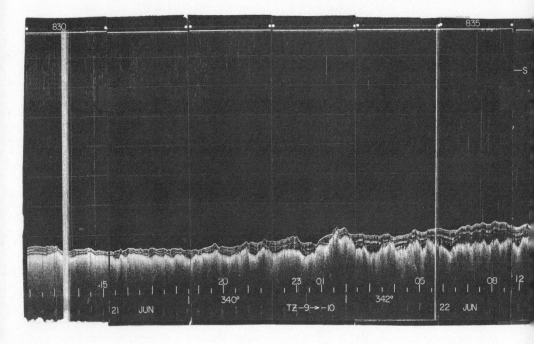

X ray of the sea floor descending beneath the Kuril Trench. Sea floor formed 110 (left) to 120 million years ago at a mid-ocean ridge near the equator starts to descend under the edge of the Asian plate. Sediments are thicker with increasing age toward the trench. Vertical faults can be seen in the sea floor already in the trench, and the echo of the sea floor continues beneath the landward (right) wall of the trench. The terraces on the right may be sediment scraped off the descending sea floor. The sea floor has bowed upward just before the lip of the trench, a characteristic feature that varies from trench to trench along with the number and severity of earthquakes and the steepness of the trench — that is, according to the ease or difficulty of underthrusting.

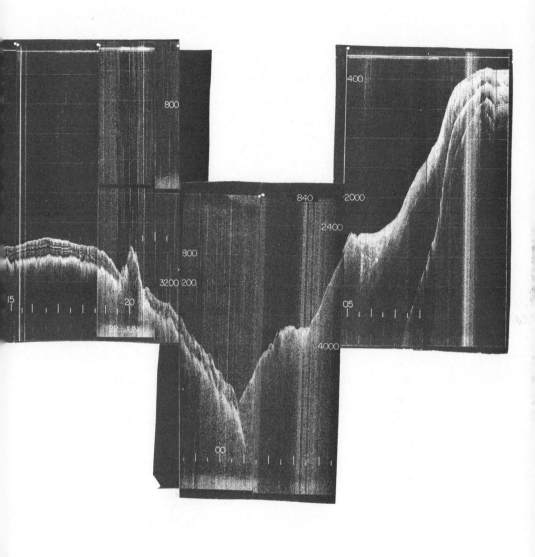

differences . . . are gross and evident even to the casual observer."

Later, Maurice Ewing and Dennis Hayes looked at profiler records of the trenches and found that they could actually see, in some, crumpled sediments apparently scraped from pieces of sea floor as they were pushed or pulled down there. How and where to dispose of as much crust as it created had been sea-floor spreading's greatest vulnerability. Ewing says: "We'd had from the start the idea the crust was pulling apart. When we found the rift, I wrote Vening Meinesz, who made the first marine gravity surveys and who had thought there was convection in the earth, that this proves it. What we couldn't see was how crust was gotten rid of again. It's hard to see why. I had repeatedly hammered the table and said, 'These trenches are empty.' Well, that's where the steel hits the rolling mill, and we can see where the sediments are scraped off against the inner wall as the crust dips under." New questions arose, some rather curious. Opdyke's group found species becoming extinct during reversals, time after time, far more than at other moments. A geologist proposed that the species were wiped out by cosmic rays when the magnetic field reached zero during a reversal. But eminent astronomers promptly, and rather acidly, offered figures to show that the number of cosmic rays which the magnetic field prevents from reaching creatures at sea level (much less below it) is quite negligible. Then a graduate student at Lamont, Billy P. Glass, found tektites at several reversals. Tektites are supposed to be fragments of comets that grazed the earth. Did the body that made the tektites make the reversal and kill the species? Neil Opdyke and James Hays of Lamont have shown the extinctions undoubtedly occur, and undoubtedly along with reversals of the magnetic field (and occasionally tektites); the connection has been thoroughly established, just not explained. After a geophysical meeting in Moscow in June 1966, before the work was published, Bruce Heezen remarked to a press conference that magnetic epochs have lasted nearly a million years; that the present epoch has lasted nearly a million years; that numerous species have become extinct at such times; and that the magnetic field has been weakening steadily for the last one hundred and fifty years and will reach zero, at its present rate, in a hundred years more.

But the confirmation by profiler records, and especially by earthquake waves, of the story already read in magnetic anomalies was another striking one of the identities that excite a scientist and that have been so lacking in geology. Lamont changed almost overnight from the most dangerous enemy of continental drift to its most fruitful advocate. And it had over half the magnetic and profiler data on the ocean basins and eighty percent of the cores waiting to be examined for evidence of sea-floor spreading. After a visit a geologist wrote Ewing: "I see why it was that your geophysicists were so largely responsible for the explosive development of plate tectonic data and concepts." At a meeting at Lamont, the geophysicist S. K. Runcorn, an advocate of drift for over a decade, remarked, genially, "I feel like a Christian visiting Rome after the conversion of Constantine."

Twelve

Global Tectonics

"IT TOOK US A LONG TIME to work out the implications of what we had just done," says Walter Pitman. "To people who had essentially rejected an idea for years, as we had rejected sea-floor spreading and continental drift, every ramification of it came hard. We had to get used to the concepts we were dealing with, and I had been so busy with anomalies that even when I had to defend my thesis I didn't know any of the arguments about continental drift. But Opdyke knew what it all meant; he knew damn well. When I first spoke on *Eltanin-19* at the American Geophysical Union meeting in spring of 1966, he went around telling people — Runcorn, Cox — that they'd see something to blow their minds. Not many people came — the abstract was so dull — but those that did were

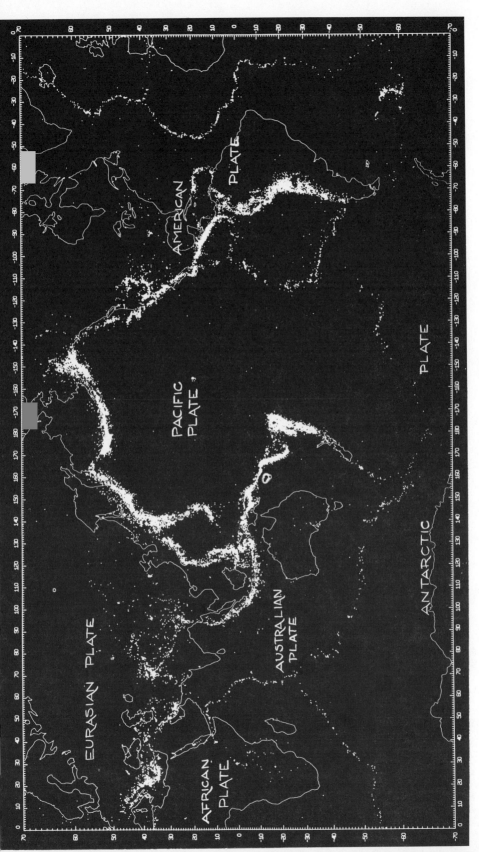

Locations of all earthquakes from 1961 to 1967, outlining the boundaries of the moving plates. Reversed for clarity.

convinced. That was the catalyst. Then NASA sponsored a meeting in November at Columbia, and that broke down all the resistance. Sykes and Oliver and Isacks had their evidence, and we presented all of ours for the first time. One man who had been violently against continental drift just got up and walked out. But I remember that Menard from Scripps, who had opposed it, sat and looked at *Eltanin*-19, didn't say anything, just looked and looked and looked. At the next AGU meeting, when we gave our four papers on the complete magnetic pattern, it was a question of breathing room. There was a special symposium on sea-floor spreading and nearly seventy papers on magnetics and sea-floor spreading, where only Opdyke and I had spoken the year before. There was no opposition." During the previous summer no one else knew what was happening at Lamont; by January the papers for the annual meeting had been submitted to the AGU. One of the begetters of sea-floor spreading, Robert Dietz, attributes "the wholesale, overnight conversion of American earth scientists to continental drift" to that meeting. By the following autumn, Professor Tuzo Wilson was sounding elegaic: "It is only ten years since Ewing and Heezen recognized that the Mid-Ocean ridges are continuous and of great size," he told a symposium in Zurich.

Several writers proposed that the spreading sea floor moves as a solid plate rotating around some point on the earth's surface — a piece of basic solid geometry. Xavier LePichon showed that the earth's surface is composed of seven large plates moving in different directions. The Americas have recently joined into one plate moving west. Most of the Pacific is moving east; and the Indian Ocean, Australia, India and Saudi Philippines. Antarctica and much of the surrounding ocean floor are stationary; Africa is moving northeast; Eurasia is moving east; and the Indian Ocean, Australia, India and Saudi Arabia are moving north. A map compiled at Lamont — by

Muawia Barazangi and James Dorman — of the locations of all the earthquakes in the world from 1963 to 1969 seems designed to depict LePichon's plates. Again, nature has been startlingly unsubtle. The earthquakes sharply define the seven large plates, plus a few small ones in the chinks. There is a certain cold majesty, like the slow turn of a galaxy, in the way the earthquake centers outline the motions of enormous slabs of the earth's surface. They run in spidery lines along the fracture zones and the crest of the Mid-Ocean Ridge. There are thick swarms of them at the trenches and mountain belts where plates come together and one plate is, in current parlance, subducted. They descend under the trenches. The earthquakes are nearly all at the plate borders; the plates are white space. By rotating the plates, like the leaves of an Astrodome, when he had found their poles, LePichon was able to translate the rate of sea-floor spreading at the Mid-Ocean Ridge into the amount of destruction at the trenches for different plates (which often meet the trenches obliquely). Unencumbered sea floor moves fastest, it appears. At New Guinea, the Pacific and Indian plates come together at eleven centimeters a year, while South America overrides the Pacific at only six. Out of the pendant zones of earthquakes below the trenches, Isacks, Oliver and Sykes wrung evidence confirming LePichon's predictions. They summed up all the earthquake evidence in an influential paper they called "Seismology and the New Global Tectonics."

The idea of rigid and rotating plates is so central that "sea-floor spreading" has become a department of "plate tectonics." The rotation of a rigid plate about an axis accounts for the different amounts, as well as kinds, of geological activity at the Ridge, the fracture zones, the trenches — the three kinds of plate border — and in the interior of the plate. The plate borders follow their own peculiar logic, transforming from one into another, docilely following geographical boundaries only suddenly to shy off elsewhere, bisecting dry land and sea floor only to enclose some of each with the other — complications which were part of the original implausibility of continental drift and sea-floor spreading. All parts of the fabric are related and affect one another. Spreading on the East Pacific Rise is matched by sliding along the San Andreas Fault and balanced by consumption of the other end of the plate at Alaska, under which it descends, via the Aleutian Trench, with shattering earthquakes. Periodically, as the two plates disengage in an earthquake, a section of coastline pops up several feet; it happened near Anchorage on Good Friday 1964. The meeting of the continent and the Pacific floor is tied also to spreading at the Mid-Atlantic Ridge; volcanism along the west coast has increased when the rate of spreading increased in the Atlantic. As the Pacific plate moves northwest, it seems to carry volcanoes away from a source of magma in the mantle; they become dormant and new ones form, producing the Hawaiian chain, which is oldest to the northwest and active only to the southeast. Cycles of activity in the Alps, and other mountains, correspond with cycles in sea-floor spreading and can be matched by date. When two irresistible objects like India and Asia meet head-on, the pattern readjusts all over the world. "It would be difficult to overstate the success of these ideas in bringing together the different disciplines that constitute earth sciences," a committee of geologists reported recently. ". . . a concept comparable to that of the Bohr atom in its simplicity, elegance, and ability to explain a wide range of diverse observations."

When Pitman and his colleagues had the magnetic pattern in the oceans decoded, they had also the tale of continental drift, the one Wegener had tried to tell. The diaspora began two hundred million years ago. There was then a single, jumbo continent, which had assembled out of smaller continents in a former convulsion of sea-floor spreading. Heirtzler, G. O. Dickson, E. M. Herron, Pitman and LePichon described the breakup in the *Journal of Geophysical Research*:

> In lowermost Cretaceous (140 million years ago), [Africa–South America] began to separate from India, Australia and Antarctica along what is now the southwest branch of the [Indian Ocean] ridge. The opening in the continent propagated clockwise around Africa, starting in Upper Jurassic (170 million years ago), near the horn of Africa, and reaching the Atlantic in Aptian times (110–120 million years ago). . . . The split in the South American and African continents may have begun almost simultaneously with the breakaway of the African–South American block. Thus the Argentine and Cape basins were born. . . .
>
> [A] second phase of spreading involved the further separation of South America from Africa and the northward movement of India. . . . The spreading on the west side of the mid-Atlantic ridge was smooth and continuous, and South America was able to drift westward over what was then a large Pacific Ocean.

India forsook Australia and Antarctica and went tearing off north, and after a mere seventy million years had run into Asia. After that collision, forty million years ago, there was a lurch and then a readjustment in the pattern of sea-floor spreading. Australia split from Antarctica while India lost momentum against Asia. New Zealand separated from Antarctica at a rapid clip and in latter years has been catching up.

Spreading in the North Atlantic has been slow and hard, and Walter Pitman and Manik Talwani were three years deciphering the pattern. They had it complete by spring of 1972. (Pitman now is in charge of marine magnetics at Lamont; Talwani ran gravity operations and became director in 1973, following Ewing's departure.) North America, Africa, and Europe did

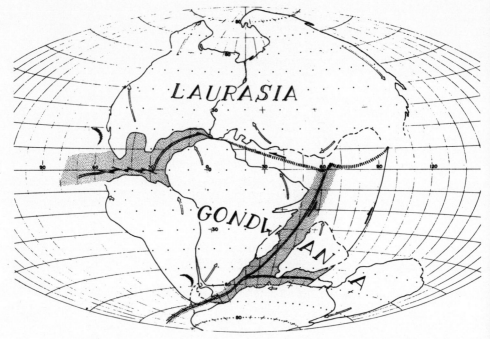

The breakup of the continents during the Mesozoic. Stippling shows new sea floor created in each stage.

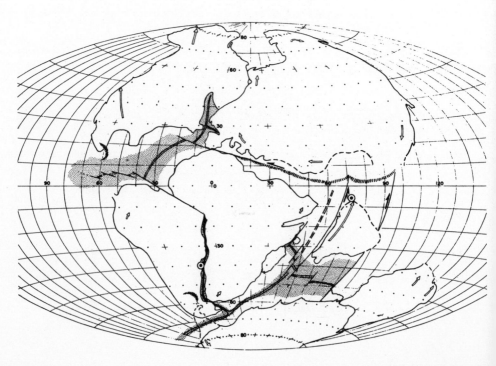

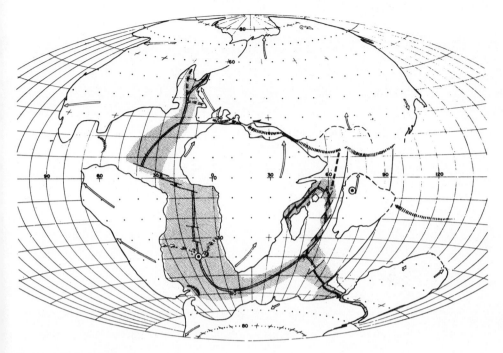

65 MILLION YEARS AGO

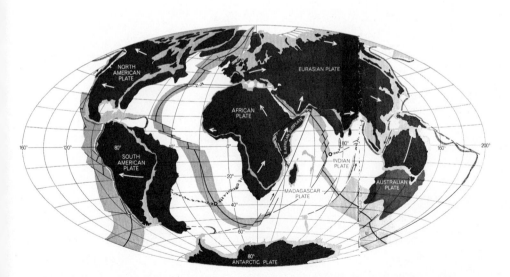

Present sea-floor spreading would produce a world like this in fifty million years. California west of the San Andreas fault has become a Canadian island, poised to carry Los Angeles into the Aleutian Trench. The Bay of Biscay and the Mediterranean have almost disappeared; the Persian Gulf has vanished. Ultimately, the Pacific may vanish, and the Americas be sutured to Asia.

not leave each other at once. Africa went (or was left) first, 180 million years ago. Dakar then would have been near Jacksonville, Las Palmas near Boston, Tangier around Cape Breton, France off Newfoundland, and England by Labrador and Greenland, there being no Labrador Sea yet. The Atlantic between North America and Africa opened rapidly for almost a hundred million years. Africa scraped past the southern edge of Europe, even for a while dragging Spain ignominiously along, opening the Bay of Biscay. Eighty-one million years ago, Europe and Greenland began to move, opening the Labrador Sea; but after twenty-five or thirty million years the center of spreading moved east of Greenland, and it was left behind. Europe traveled fast and was gaining on Africa until fifty million years ago. Since then, they have kept pace, but Africa has steadily crowded north against Europe. The motion has not been easy on the soft underbelly of Europe, which has been repeatedly squeezed. "It's one of the ironies of geology that the Alps, where classical geology began, have the most complicated sequence of orogenies in the world," said Pitman after describing the relative motions between the two eastbound continents. Spain and other small fragments have been traded back and forth. Italy belongs to the African plate (not Africa to Italy). The Alps are where the African plate is thrusting against the European — the plates are being, as it is called now, sutured. Beneath the eastern Mediterranean, with a seismic profiler, the sediments can be seen upheaved and crumpled in the embryonic stages of mountains. Africa will presently collide with Greece.

In the mid-sixties, Lamont chartered an oil company drilling ship to drill a few deep sediment samples from the continental margin, a project that became the JOIDES, or Joint Oceanographic Institutions Deep Earth Sampling, program. Since 1968, a ship capable of drilling even at the depths of the ocean basins has been doing so under charter to the National Science

Foundation, and under the direction of Scripps, Lamont, and Woods Hole. In its first year, in the South Atlantic, it drilled a series of holes to the basement between the Ridge and the continental margin. The sediment immediately above the basement got older with greater distance from the rift, and the ages of the samples were those predicted from the magnetic anomaly pattern. Even without such tangible evidence, it seems certain now that all the ocean floors are less than two hundred million years old. This is the age of the first part of the Atlantic to form, the ocean floor immediately off the east coast, both by drilled samples and by extrapolation from the magnetic pattern. There is slight hope that a scrap of older stuff may have escaped recycling and remain still to be found in some cranny of the world.

Several people, at Lamont and elsewhere, have argued that the Caribbean, a small independent plate that the Atlantic dips under at the Antilles, is a remnant of the ancient Pacific that slipped between North and South America before they were joined; many things make the idea an attractive one, but drilling appears to leave the Caribbean too young. The Caribbean remains puzzling; there is no room for it or its islands, or Central America, in reconstructions of the continents before drift. Recent spreading of the Indian Ocean has erased any direct evidence of precisely where India came from; several writers put it and Madagascar next to Somalia, while others put them both down by Mozambique, flanked by Australia and Antarctica. The Pacific too shows few signs of its past; half the Pacific sea floor has been swallowed up beneath the Americas. Japan may be the suturing of several island arcs; a small portion of Siberia may have drifted over from the American Southwest. The magnetic pattern in the Pacific stops several thousand miles from California, and short of Hawaii. The fracture zones continue on to the islands. Farther west there is yet another magnetic pattern, which correlates with anomalies in the Atlantic, between Bermuda and the East Coast. From sediment drilling in both areas, these anomalies appear to have been formed between 130 and 110 million years ago. Immediately after that cycle of spreading, however, from 110 to 85 million years ago, the earth's field did not reverse at all; none of the sea floor created in that period bears a magnetic pattern, and the early history of the Pacific can only be inferred.

But most of the recent history of the continents and oceans is stated with some assurance; the outline of global geology seems both clear and generally accepted now. There are a few dissenters, but just a few — too few, really, to keep an establishment on its toes. There have been several bizarre interpretations of the magnetic anomalies. A. A. Meyerhoff claims that the idea of a supercontinent, a Pangaea, is self-defeating in at least one respect: more land would be far from the sea and from the moisture necessary to form the Gondwana icecaps, and coal. Sir Harold Jeffreys, a great expert on such matters, maintains that the mantle will not flow as at least some of it must if sea-floor spreading occurs. Cambrian trilobites were dredged, once, in mid-Atlantic near the Azores, rather like finding King Tut in Grant's Tomb. To all of which the confirmed drifter might, echoing Galileo, murmur, *E pur se muove.* Most men seem now to regard spreading not as a theory, or even a hypothesis, but a fact. Even Ewing, after years as devil's advocate (for which he is vividly remembered), called it a ninety-nine percent plus fact. The implications of sea-floor spreading seem endless. Ideas for predicting earthquakes have been reversed. Serious earthquakes are now expected, not where there have been frequent earthquakes, but in adjacent areas where there have been none recently. There, for unknown reasons, instead of dissipating in small earthquakes, strain is accumulating ominously as one plate moves upon another. Several of these spots are found on the San Andreas Fault and in Alaska.

Consider also the turtle, who lives in Brazil and breeds and nests on Ascension Island in the middle of the South Atlantic Ocean. The navigation of the turtle has for some years been the admiration of biologists, who despaired of understanding it. Why does the turtle trouble to go so far to a speck of an island, and how did it first hear that the island was there? Almost certainly the turtle's habit predates the separation — or most of it — of Africa and South America. Could its trip have been a short one originally, and been lengthened by tiny amounts each year by sea-floor spreading?

Spreading has a few commercial effects. A salt mine was found in Brazil recently after similar deposits were found in the once-contiguous part of West Africa. It has even been suggested that "crustal sinks" — that is, the trenches — are a good place to dispose permanently of rubbish, especially the radioactive kind. The rift valley produces mineral deposits (at the bottom of the Red Sea, a young rift, are pools of very salty and hot water under which are several hundred feet of sediments heavy with copper, zinc, even gold); turbidity current deposits develop oil. It can be seen now that lodes on the continents were accumulated by the processes of sea-floor spreading, bits of sea floor being left behind as one plate dives under another plate. Although the greatest part of the sea floor is subducted, and the mineral deposits thus are probably largely accidental, there is a clear advantage to knowing just how such parts of the sea floor are incorporated into the continents, and commercial geologists are working on it.

Mountain-building has been a central and basically insoluble problem in geology since its earliest days; and the apparent discrepancy between the simple localized pattern of sea-floor spreading and the extensive complicated patterns of mountain

An early stage in the growth of an ocean. The continents have only been separated some ten million years. The sea is relatively shallow, with little circulation; the bottom is stagnant in places. The ridge flanks are not yet buried
in sediment, and in the rift valley at moderate depths there are valuable
mineral deposits — and very hot water. Africa lies to the right; Saudi Arabia
to the left. North is down.

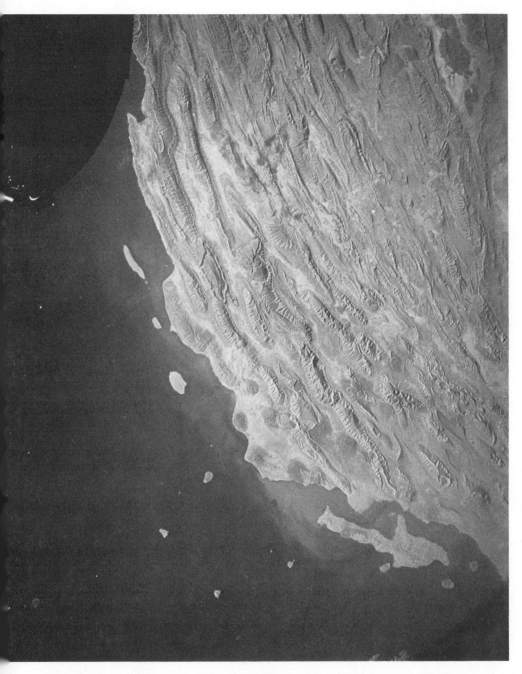

The Saudi Arabian plate, moving northeast, meets the Asian plate in Iran
along a thrust fault that crosses the upper right corner of the picture. The edge
of the Arabian plate has crumpled into long folds, which are preserved by a
low rate of erosion there. The Persian Gulf (left) is part of the Arabian plate.
Arched and folded sediments make good reservoirs for petroleum.

formation — crust is created and destroyed in zones under ten miles wide, while mountains can be seven hundred wide — has been called the chief reason geologists were slow to take sea-floor spreading seriously. Spreading does account for the energy, the raw materials, and the timetables of mountain-building. In the last year or two geologists have begun identifying in mountain systems not only transformed abyssal plain sediments, trench sediments, and ordinary abyssal sediments, but pieces of ocean crust, and even intrusions of mantle, in Cyprus and California and Newfoundland, for example. (These places already had been the sites of some mining.) Years ago, before its significance was clear, the combination of basalts, pillow lavas, chert (a marine sedimentary rock) and limestone, in successive layers, was called an ophiolite suite. The ophiolites of Cyprus, the Troodos massif, appear to be sea floor created a few tens of millions of years ago at a Ridge, in a rift, carpeted with sediment, and thrust out of water as the African plate overrode Europe. Several sites in Newfoundland are exciting because they show the same things were happening eons before today's bout of sea-floor spreading began — before even Pangaea. But of what ocean are these time-worn rocks remnants?

How pieces of the sea floor have gotten raised up into mountains is not precisely known. Some may be scraped off in the trenches, or transformed farther down, or buckled up. The fragments of a descending plate may become attached to the continent at great depth and only after hundreds of millions of years of erosion be exposed to view. (In a few cases — New Britain is one — sheets of sea floor have slid over, instead of under, the edge of another plate.) But in them can be read accounts of the growth and destruction of oceans. Mountains that are active today are the only lasting record of present sea-floor spreading; and it is likely that older mountains record the openings and closings of vanished oceans, and the continental gavottes that formed the land into Pangaea and even earlier conglomerations. As the ideas of Hutton and Smith made it possible to read one kind of history in the rocks, now those of plate tectonics translate one of continents and oceans. The Appalachian Mountains and the Caledonian Mountains of Scotland and Scandinavia were continuous before the opening of the Atlantic, and they record the opening and closing of an earlier ocean, commencing in the late pre-Cambrian. The rocks from Newfoundland to Alabama are largely paleozoic; in New York they are particularly neatly laid out from the Catskills north to the pre-Cambrian Adirondacks, succeeding each other across the land as regularly as the leaves of a book. This was the continental shelf of a widening ocean and later a marginal sea behind an island arc, shallow, rich in life like those of today, and warm and near the equator. In those days, a forerunner of North America which included northern Scandinavia and Scotland, an Africa that included Georgia, Florida and South Carolina, and Europe, had all parted from a single continent that broke up in the late pre-Cambrian. Miles of sediments accumulated under the shallows during the next several hundred million years, while life evolved the first shelled animals, coral reefs, and, toward the last, fish. There were limestones on the continental shelf, and turbidites on the continental margin. Through the Cambrian and the early Ordovician periods the ocean grew. (Africa drifted over the South Pole, and in the Sahara rocks still shows the scars of the glaciers of that time.) Then the direction of continental drift reversed and the ocean began to contract. A trench and then an island arc formed off the North American coast, where the ocean floor was consumed. Sediments from the islands covered the old continental margin, while to seaward sediment scraped off the descending ocean floor accumulated on the inner wall

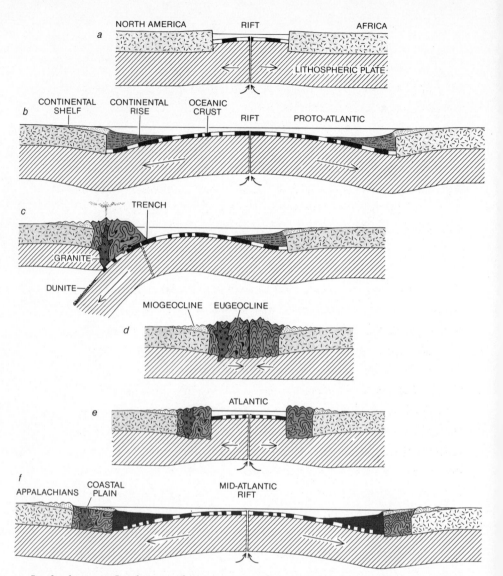

In the late pre-Cambrian, a large continent began to break up and a new ocean to form. The incipient North America then included parts of Scotland and Scandinavia. The ocean widened during the Cambrian, and sediments from the continents formed continental shelves and rises like today's. The seas teemed with life, but on the continents there was none. Almost 500 million years ago the direction of continental drift was reversed, and the ocean began to close; the sea floor was thrust under the American continental margin, raising the heavy accumulations of sediment into mountains. At the end of the Silurian, Europe collided with North America, raising a chain of alps that is now spread from Newfoundland to Norway. For a hundred million years there was peace; life crawled out onto the land. Then Africa plowed into Europe and America, raising a chain of mountains from Alabama to Poland. Erosion wore them down almost to stumps. Two hundred or so million years ago North America was on the move again. It left Scotland for the English and borrowed Georgia and Florida from West Africa. The rifting produced floods of basalt; the Hudson Palisades are some of what is left. Erosion beveled the edges of the separating continents. New accumulations of sediments built up into a continental shelf and rise.

of the trench. In the late Silurian, and during the Devonian period, the ocean was pinched out and continents collided; the activity continued through the Devonian, even as today the ramming together of India and Asia continues. First the island arc, then Europe, piled into the North American margin, making mountains from New England and Nova Scotia to Scotland and Finland; finally Africa collided, making more mountains from Alabama to Hungary. In a similar way today India's collision with Asia has destroyed an intervening ocean, and Africa is still moving on Europe, pinching out the Mediterranean. The present interprets the past and the past foreshadows the future.

The ancient mountains were at least the equal of the Alps and Himalayas. By the Mesozoic they had been eroded to a fairly level plain, and a rift was developing near the old lines of collision. North America was on the move again, borrowing most of the Confederacy from West Africa and subsequently leaving the Scottish highlands to the English. There were great outpourings of basalt, of which the Hudson Palisades are one remnant. The heat of the mantle lifted the edges of the continents and erosion thinned and beveled them. By the Jurassic the Atlantic was a narrow sea with room for little more than a rift and incipient ridge. There was little circulation in the warm waters; the sea may have dried out completely from time to time. Large deposits of salt and of minerals accumulated with the sediments. Sediments from the land buried them and made new continental margins and shelves. The beveled edges of the continents sank lower as they got farther from the rift. Fifty million years ago currents began to flow in the ocean, scouring some parts, depositing dunes in others; the Gulf Stream flowed into the Labrador Sea until the rift between Greenland and Britain was wide enough.

The Urals are another suture of former continents, the re-

233

mains of an ocean that existed until the Triassic period only shortly before Pangaea began to break up again. There are pre-Cambrian sutures in Africa, and several in northern Canada, 1,100, 1,900 and 2,800 million years old respectively, the relics it seems of hegiras and pangaeas so far away in time that they can be only dimly seen. "Continental drift goes as far back in geological time as we can see," one geologist has said, echoing Dr. Hutton. All this, however, is extrapolation, not proof. If the old rocks can be interpreted in the light of, and in the same manner as, new rocks, then this is the result. But there is some contradictory evidence, suggesting that the breakup of Pangaea was a unique catastrophe, not a routine event, and that the land had always been in one piece before. Some geologists find it completely implausible that the continents, once dispersed, would ever all find each other again, while others think it only to be expected that they would, after a time, meet halfway around the world from where they parted company. But plate tectonics accounts for many rather basic observations which for a long time left geologists embarrassed for an explanation: why the exceptional areas of both continents and oceans — mountain systems, island arcs and trenches, the Mid-Ocean Ridge — are long narrow strips; why areas of active mountain-building have not persisted, but been different from age to age; why mountains have had such a predilection for coastlines and why much of their sediment, and the forces than crumpled them, have come from seaward. The lost continent of Appalachia was Africa.

The story of the succession of former continents is only beginning to be told. Their divisions and recombinations have left their mark not only in the rocks but in the progress of life. There have been great ages of high sea level, when the continents lay almost entirely under shallow seas, and ages of great flourishing and diversification of species. There have been ages of mass extinction as well — at the end of the Permian, less than half the species of animals survived. Such worldwide, wholesale changes of the earth and life upon it probably relate, directly or indirectly, to sea-floor spreading. The suturing of continents at the end of cycles of spreading would increase competition between species, and extinction of those least able to adapt or compete. When continents broke up, species would diversify. The depth of the ocean floor is greater with greater age, and the ridges with the slowest spreading have the steepest flanks. In 1972, Pitman and Larson at Lamont discovered that the speed of sea-floor spreading has increased enormously during periods of unchanging magnetic polarity. Large volumes of young sea floor — Pitman and Hays wrote in 1974 — would raise sea level by hundreds of yards; and fast spreading would mean also fast sea-floor consumption and destruction under the mountains at far ends of the plates. The last period of constant polarity, from 110 to 85 millions years ago, when the speed of sea-floor spreading suddenly tripled, was also a period when the oceans spread over the land in great shallow inland seas; at the same time, judging from the vast bodies of granite left from that period — in the Sierras and elsewhere — there were paroxysms at the plate boundaries. And at the end there were mass extinctions. Ridges subsided, sea level dropped, the shallow seas — where the richest assortment of creatures has always lived — dried out. Most species died. The catastrophe was worldwide and far more relentless than any flooding.

A revolution has taken place, but much that is discussed remains fairly speculative, while even some basic physical observations, like the composition of the Ridge and its location in all the oceans, and the extent and positions of fracture zones, are interlarded as most knowledge is with inference and assumption. As with an election forecast, the overall result is known — with virtual certainty — before the composition of the electorate, or the how and the why of the result have been analyzed. If the pattern of the sea floor over thousands of miles or millions of years is clear, what is happening day by day or year by year in the Ridge is not so clear, nor is exactly where or how it is happening. What is a Ridge, a cause or an effect? And how, especially, does one begin? The African rift valleys are called the beginnings of continental breakup, except by those who call them an aborted case. (As evidence that abortions occur, there appears to be in the middle of the United States a pre-Cambrian rift valley from Lake Superior to Kansas. Newer sediments have filled it, but it seems to have reached a later development than the African rifts, more like the Red Sea, and it is interesting that some of the great sedimentary deposits of iron of the Great Lakes region are associated with it.) Similarly, the course of suturing of plates, and mountain making, are more easily described in generalities than particulars. Why and how it all occurs is only somewhat clearer than before anyone saw *Eltanin*-19. The expanding earth idea has been revived from time to time. One writer has suggested the continents have sufficient gravitational attraction for one another to move themselves at the rates observed. The coriolis force — the effect of the earth's rotation — has been implicated. The huge weight of the Gondwanaland icecap has been mentioned as a cause of continental rupture. It is not at all obvious why the plates should have odd shapes and apparently haphazard borders, as they do, nor is how or where their motion is transferred to them. While convection in the mantle has been an enduring candidate, no one is able to say with much detail or precision how convection can occur in the mantle, nor specify the size, shape, or composition of what it is that convects — if the mantle convects. (To find the structure of the

earth from earthquake waves, the oldest and thus far most pro-
ductive way has been likened to trying to deduce the structure
of a piano from the noise it makes falling down a flight of stairs.)
Just now, mantle "plumes," or "hot spots," one of which is supposed to be responsible for the Hawaiian Islands, are in fashion
as the force that drives sea-floor spreading. There appear to be
several dozen hot spots around the globe — Iceland is thought
to be another — generally identifiable by volcanic islands that
have formed above them and then been carried away in long
chains by sea-floor spreading. Beneath each hot spot there is
thought to be a mantle plume — a rising column of extra-hot
rock. The plume, perhaps, is what stimulates the crust to
move. No one knows much about plumes — what they are,
why they are where they are, how they could start the whole
crust moving, not to mention whether they really exist — but
thus far the hot spots are the only sign that anyone has found
of something in the mantle that is moving in the right direction. The mantle, once a thick, conveniently homogenous
layer between crust and core, has been parceled out. Geologists speak now of a lithosphere of seventy kilometers, of
which part is former mantle and part crust (some think the
continents are the only crust, that the sea floor is the chilled top
of the mantle). It is the lithosphere that charges about the
surface of the earth. There is an asthenosphere, which is soft
and flowing, under the moving lithosphere, and beneath that a
mesosphere which is hard enough that descending lithosphere
seems to bottom against it. At first it looked as though the
spreading sea floor is pushed apart at the rift valley by rising
magma; now opinion favors the opposite view, that the plate
runs downhill into the trench, pulling its tail parts after it. The
Ridge is a sort of welt lifted by the heat of stuff rising in the
mantle; the sea floor slides down the slope of its flanks. There is
tension at the Ridge and new stuff works its way in, a line of be-

havior followed by water in crevasses. The far end of the plate, descending into the mantle, should pull harder as it gets deeper, until it hits the mesosphere; and sea-floor spreading does speed up, and slow down, over millions of years. This is no ordinary convection, though there is a rise and a return; it is far from the convection currents that originally were supposed to carry the spreading sea floor on their backs. "I avoid the term 'convection cells' as much as people in the Middle Ages avoided the term 'devil,'" says one physicist. "All one can say is that material goes down some place and comes up another place."

Talk of sea-floor spreading has assumed crust and mantle are constantly mixed by the process in an endless cycle of regeneration — though sometimes the talk is of the mantle rejuvenating the crust and sometimes of the crust rejuvenating the mantle. But another idea is possible, suggested in *Science* in 1971, that the asthenosphere is the source of crust and the mesosphere its midden, that crust congeals out of the one and travels until it adds itself to the other. The sizes and proportion of the two zones accord with three billion years of sea-floor spreading in the past — and with potential for a billion more before the zone of soft flowing material is used up and the hard zone extends all the way up from the earth's core to the crust. Thereafter there would be no more mountain-building, only erosion. Often drifters have spoken of mantle activity as though it existed to create sea-floor spreading; this is not likely to be the case. There are intriguing connections between earthquakes and irregularities in the earth's rotation, which probably have a common cause. Magnetic field reversals and cycles of reversals are implicated with them, and it seems odd, too, that the poles of sea-floor spreading cluster near the magnetic poles. Great ripples seem to run across the continents over hundreds of millions of years. All point toward mysterious goings-on in the mantle. While plate tectonics is a beautifully simple notion, the causes of it and the changes rung on it are not; plate margins and sutures are likely to be the problem of the next five or ten years in geophysics and geology, and the mantle the problem of the next twenty. "Nothing is going to be as simple as plate tectonics now makes it seem," Ewing said the other day, "just as no atom is exactly like the Bohr atom, but that was an idea capable of sufficient refinement. There are things going on in the earth that are just as cunning as this magnetic pattern." Many of these things will be found out at sea, just as the magnetic pattern was. The ships of Lamont, Woods Hole and Scripps are still looking. Ewing's new Marine Sciences Institute, of the University of Texas system, has a new ship in Galveston, the *Ida Green*, which he proposed to concentrate on the Caribbean, whose place in the fabric of plate tectonics is still mysterious; there is continuing work on the seismology of the earth, and the moon, and a project afoot to measure the motion of the continents by laser beam. It seems unlikely that geologists will soon revert to the attitude that everything worth knowing is known, which Ewing said was one of the chief reasons he thought the field would be rewarding to work in. For the first time it is possible to see, broadly, the life of the earth, how it sustains and renews itself, and the growth and evolution of its surface which rises up as often as it is eroded down.

Bibliography
and Works Consulted

Suggested reading

Continents Adrift, readings from *Scientific American*, intro. J. Tuzo Wilson, W. H. Freeman, San Francisco. Articles for laymen by some of the men who established sea-floor spreading. (Similarly: When the Mediterranean Dried Up, K. J. Hsü, *Scientific American*, December, 1972, pp. 26–36; and Plate Tectonics and Mineral Resources, P. A. Rona, *Scientific American*, July 1973, pp. 86–96.)

Debate About the Earth, Takeuchi, Uyeda and Kanamori, 1967: W. H. Freeman, San Francisco. Drifters and non-drifters up to the mid-sixties.

The Deep and the Past, D. B. Ericson and Goesta Wollin, 1964: Knopf, New York. Sampling the bottom at sea, and how the samples are deciphered into a history of the earth during the rise of man.

The Edge of an Unfamiliar World, A History of Oceanography, Susan Schlee, 1973: E. P. Dutton, New York. Mostly American, mostly before 1950, with a chapter on later developments.

241

The Face of the Deep, B. C. Heezen and C. D. Hollister, 1971: Oxford U. Press, New York. The abyss portrayed; 600 superb photographs.

Geology Illustrated, J. S. Shelton, 1966: W. H. Freeman, San Francisco. Beautiful pictures again; on land, this time.

The New World of the Oceans, Men and Oceanography, D. Behrman, 1969: Little, Brown, Boston. Scripps and Woods Hole these days; and a little geology at Lamont.

The Ocean, a Scientific American Book, W. H. Freeman and Co., San Francisco.

A Revolution in the Earth Sciences, From Continental Drift to Plate Tectonics, A. A. Hallam, 1973: Oxford U. Press, New York. Good history of ideas. No personalities.

The Sea Around Us, Rachel Carson, 1951: Oxford U. Press, New York.

Understanding the Earth, I. G. Gass, P. J. Smith and R. C. L. Wilson, 1971: MIT Press, Cambridge. A comprehensive gaggle of articles by earth scientists, originally for The Open University in England.

Original research, a little more difficult

Adventures in Earth History, Original Selections from Steno to the Present, Preston Cloud, 1970: W. H. Freeman, San Francisco. Historical geology.

Plate Tectonics and Geomagnetic Reversals, Allan Cox, ed., 1973: W. H. Freeman, San Francisco. Most of the classic papers. Good background by Cox.

The Sea, v. 3, M. N. Hill, ed., 1963: Interscience, New York. The sea floor as it was then known; all the postwar discoveries.

Works consulted

Chapter One
The Floor of the Sea

Analysis of a feed-back controlled seismometer, G. H. Sutton and G. V. Latham, *Jour. Geophys. Res.*, v. 69, pp. 3865–3882, 1964.

The Apollo passive seismic experiment, G. Latham, M. Ewing, F. Press, and G. Sutton, *Science*, v. 165, pp. 241–250, 1969.

Continuous gravity measurements on a surface ship, J. L. Worzel, *Jour. Geophys. Res.*, v. 64, pp. 1299–1315, 1959.

Dewbows by moonlight, M. Ewing, *Science*, v. 63, pp. 527–528, 1926.

Gravity at sea, J. L. Worzel and J. C. Harrison, *The Sea*, v. 3, 1963: Interscience, N.Y., pp. 134–174.

Chapter Two
The Chief Scientist Dreamed of Cervantes

Deep sea measurements without wires or cables, M. Ewing and A. C. Vine, Amer. Geophys. Union, *Trans.*, 1938, pp. 248–251.

Early development of ocean-bottom photography. . . . , M. Ewing, J. L. Worzel, and A. C. Vine, *Deep Sea Photography*, 1967: Johns Hopkins Press, pp. 13–41.

Geophysical investigations in the emerged and submerged Atlantic coastal plain. Part I: Method and result, M. Ewing, A. P. Crary, H. M. Rutherford, *Geol. Soc. Amer. Bull.*, v. 48, pp. 753–802, 1937. Part II: Barnegat Bay, N.J., section, M. Ewing, G. P. Wollard, and A. C. Vine, *Geol. Soc. Amer. Bull.*, v. 50, pp. 257–296, 1939. Part IV: Cape May, N.J., section, *ibid.*, v. 51, pp. 1821–1840, 1940.

Gravity measurements at sea, J. L. Worzel and M. Ewing, Amer. Geophys. Union, *Trans.*, v. 31, pp. 917–923, 1950.

Photography of the ocean bottom, M. Ewing, A. C. Vine, and J. L. Worzel, *Jour. Opt. Soc. Amer.*, v. 36, pp. 305–321, 1946.

Sub-surface constitution of Bikini Atoll as indicated by a seismic-refraction survey, M. B. Dobrin, B. Perkins, Jr., and B. L. Snavely, *Geol. Soc. Amer. Bull.*, v. 60, pp. 807–829, 1949.

Chapter Three
Lifeless Depths, Living Fossils, Lost Continents

The Depths of the Sea, C. Wyville Thomson, 1874: Macmillan, London.

Founders of Oceanography, W. A. Herdman, 1923: Arnold, London.

Notes by a Naturalist on the "Challenger," H. N. Moseley, 1879: Macmillan, London.

The Physical Geography of the Sea, M. F. Maury, 1858: Harper and Bros., New York.

Chapter Four
A Scene the Most Rugged, Grand, and Imposing

Elementary theory of seismic refraction and reflection measurements, J. I. Ewing, *The Sea*, v. 3, 1963: Interscience, New York, pp. 3–19.

North Atlantic hydrography and the Mid-Atlantic Ridge, I. Tolstoy and M. Ewing, *Geol. Soc. Amer. Bull.*, v. 60, pp. 1527–1540, 1949.

Seismic investigations in great ocean depths, M. Ewing, Assoc. Oceanog. Phys., *Proc.-Verb.*, v. 5, pp. 135–136, 1952.

Seismic Reflections from beneath the Ocean Floor, J. B. Hersey and M. Ewing, *Trans. Amer. Geophys. Union*, v. 30, pp. 5–14, 1949.

Seismic shooting at sea, M. Ewing and L. Engel, *Sci. Amer.*, v. 206, no. 5, pp. 116–126, 1962.

Submarine topography in the North Atlantic, I. Tolstoy, *Geol. Soc. Amer. Bull.*, v. 62, pp. 441–450, 1951.

Chapter Five
Torrey Cliff

The Atlantic Ocean Basin, M. Ewing, *Bull. Am. Mus. Nat. Hist.*, v. 99, no. 3, pp. 87–91, 1952.

Atmospheric waves from nuclear explosions, W. L. Donn and M. Ewing, *Jour. Geophys. Res.*, v. 67, pp. 1855–1866, 1962.

The crust and mantle of the earth, M. Ewing, Amer. Geophys. Union, *Geophys. Mon. 2*, pp. 186–189, 1958.

Crustal structure and surface-wave dispersion, M. Ewing and F. Press, *Seism. Soc. Amer. Bull.*, v. 40, pp. 271–280, 1950.

Crustal structure and surface-wave dispersion, Part iv: Atlantic and Pacific Ocean basins, J. Oliver, M. Ewing, and F. Press, *Geol. Soc. Amer. Bull.*, v. 66, pp. 913–946, 1955.

Earthquake surface waves and crustal structure, F. Press and M. Ewing, *Geol. Soc. Amer. Spec. Paper 62*, pp. 51–60, 1955.

An explanation of the Lake Michigan wave of 26 June, 1954, M. Ewing, F. Press, and W. L. Donn, *Science*, v. 120, pp. 684–686, 1954.

Exploration of sub-oceanic structure by the use of seismic surface waves, J. Oliver and J. Dorman, *The Sea*, v. 3, 1963: Interscience, New York, pp. 110–133.

Geophysical contrasts between continents and ocean basins, M. Ewing and F. Press, *Geol. Soc. Amer. Spec. Paper*, v. 62, pp. 1–5, 1955.

The Lamont Geological Observatory, G. W. Gray, *Sci. Amer.*, Dec. 1956, pp. 83–94.

Long period seismic waves from nuclear explosions in various environments, J. Oliver, P. Pomeroy, and M. Ewing, *Science*, v. 131, pp. 1804–1805, 1960.

Mantle Rayleigh waves from the Kamchatka earthquake of November 4, 1952, M. Ewing and F. Press, *Seism. Soc. Amer. Bull.*, v. 44, pp. 471–479, 1954.

Notes on Surface Waves, M. Ewing and F. Press, *Ann. N. Y. Acad. Sci.*, v. 51, pp. 453–462, 1949.

The Sea Around Us, R. Carson, 1951: Oxford U. Press, New York.

Seismic measurements in ocean basins, M. Ewing and F. Press, *Jour. Mar. Res.*, v. 14, pp. 417–422, 1955.

Seismic refraction measurements in the Atlantic Ocean Basin. Part I: M. Ewing, J. L. Worzel, J. B. Hersey, F. Press and G. R. Hamilton, *Seism. Soc. Amer. Bull.*, v. 40, pp. 233–242, 1950; Part IV: Bermuda, Bermuda Rise, and Nares Basin, C. B. Officer, M. Ewing, and P. C. Wuenschel, *Geol. Soc. Amer. Bull.*, v. 63, pp. 777–808, 1952.

Seismic waves coupled to sonic booms, J. Oliver and B. Isacks, *Geophysics*, v. 27, no. 4, pp. 528–530, 1962.

Shape and structure of the ocean basins, M. Ewing and M. Landisman, *Oceanography*, Amer. Assoc. Adv. Sci., 1961, pp. 3–38.

Trans-Atlantic profile of total magnetic intensity and topography, Dakar to Barbados, B. C. Heezen, M. Ewing, and E. T. Miller, *Deep-sea Res.*, v. 1, pp. 25–33, 1952.

Chapter Six
That Man Is Taking Too Many Cores

Abyssal plains, B. C. Heezen and A. S. Laughton, *The Sea*, v. 3, 1963: Interscience, New York, pp. 312–364.

Alice in Greywackeland, P. D. Krynine, *Jnl. Pal.*, v. 30, pp. 1003–1004, 1956.

Atlantic deep-sea sediment cores, D. B. Ericson, M. Ewing, G. Wollin, and B. C. Heezen, *Geol. Soc. Amer. Bull.*, v. 72, pp. 193–285, 1961.

Coiling direction of globigerina pachyderma as a climatic index, D. B. Ericson, *Science*, v. 130, pp. 219–220, 1959.

Coiling directions of Globorotalia truncatilinoides in deep-sea cores, D. B. Ericson, G. Wollin, J. Wollin, *Deep-sea Res.*, v. 2, pp. 152–158, 1954.

Cross-correlation of deep-sea sediment cores. . . . , D. B. Ericson, *The Sea*, v. 3, 1963: Interscience, New York, pp. 832–842.

The deep sea and early man, M. Ewing, D. B. Ericson, A. W. Bally, and G. Wollin, *Quaternaria*, v. I, pp. 145–168, 1954.

Deep-sea sands and submarine canyons, D. B. Ericson, M. Ewing and B. C. Heezen, *Geol. Soc. Amer. Bull.*, v. 62, pp. 961–965, 1951.

Essais de comparaison entre les turbidites modernes et le flysch, W. D. Nesteroff and B. C. Heezen, *Revue de Geographie Phys. et de Geologie Dynam.*, v. 5, fasc. 2, pp. 113–125, 1963.

Estimated size of the Grand Banks turbidity current, Ph. H. Kuenen, *Amer. Jnl. Sci.*, v. 250, pp. 874–887, 1952.

Experiments in connection with Daly's hypothesis, Ph. H. Kuenen, *Leidsche Geologische Mededeelingen*, deel 8, pp. 327–351, 1938.

Exploration of the deep-sea floor, M. Ewing, D. B. Ericson, D. C. Heezen, J. L. Worzel, and G. Wollin, *Quaternaria*, v. 1, pp. 145–168, 1954.

Exploration of the northwest Atlantic mid-ocean canyon, M. Ewing, B. C. Heezen, D. E. Ericson, J. Northrop, and J. Dorman, *Geol. Soc. Amer. Bull.*, v. 64, pp. 865–868, 1953.

The Floor of the Ocean, R. A. Daly, 1941: University of North Carolina Press, Chapel Hill.

Further evidence for a turbidity current following the 1929 Grand Banks earthquake, B. C. Heezen, D. B. Ericson, and M. Ewing, *Deep-sea Res.*, v. 1, pp. 193–202, 1954.

The Grand Banks slump, B. C. Heezen and C. L. Drake, *Bull. Amer. Assoc. Petr. Geol.*, v. 48, pp. 221–225, 1964.

The influence of submarine turbidity currents on abyssal productivity, B. C. Heezen, M. Ewing, and R. J. Menzies, *Oikos*, v. 6, pp. 170–182, 1955.

Late-Pleistocene climates and deep-sea sediments, D. B. Ericson, W. S. Broecker, J. L. Kulp, and G. Wollin, *Science*, v. 124, pp. 385–389, 1956.

Modern graywacke-type sands, C. D. Hollister and B. C. Heezen, *Science*, v. 146, pp. 1573–1574, 1964.

North Atlantic deep-sea sediments and submarine canyons, D. B. Ericson, *Trans.*, N.Y. Acad. Sci., Ser. II, v. 15, pp. 50–53, 1952.

The origin of submarine canyons, B. C. Heezen, *Sci. Amer.*, Aug. 1956, pp. 36–41.

Our shrinking globe, K. K. Landes, *Geol. Soc. Amer. Bull.*, v. 63, pp. 225–240, 1952.

Physical properties of marine sediments, J. E. Nafe and C. L. Drake, *The Sea*, v. 3, 1963: Interscience, New York, pp. 794–815.

The Pleistocene epoch in deep-sea sediments, D. B. Ericson, M. Ewing and G. Wollin, *Science*, v. 145, pp. 723–732, 1964.

Puerto Rico Trench topographic and geophysical data, M. Ewing and B. C. Heezen, *Geol. Soc. Amer. Spec. Paper 62*, pp. 255–268, 1955.

Reconnaissance survey of the abyssal plain south of Newfoundland, B. C. Heezen, M. Ewing and D. B. Ericson, *Deep-sea Res.*, v. 2, pp. 122–133, 1954.

Remarks on the Grand Banks turbidity current, B. Kullenberg, *Deep-sea Res.*, v. 1, pp. 208–210, 1954.

Sediment deposition in the deep Atlantic, D. B. Ericson, M. Ewing, B. C. Heezen and G. Wollin, *Geol. Soc. Amer. Spec. Paper 62*, pp. 205–220, 1955.

Submarine erosion, a discussion of recent papers, F. Shepard, *GSA Bull.*, v. 62, pp. 1416–1420, 1951.

Submarine topography in the North Atlantic, B. C. Heezen, M. Ewing, and D. B. Ericson, *Geol. Soc. Amer. Bull.*, v. 62, pp. 1407–1417, 1951.

Textural evidence for deposition of many western north Atlantic deep-sea sands by ocean-bottom currents rather than turbidity currents, J. F. Hubert, *Jour. Geol.*, v. 72, pp. 757–785, 1964.

Turbidity currents, B. C. Heezen, *The Sea*, v. 3, 1963: Interscience, New York, pp. 742–775.

Turbidity currents and graywackes, F. J. Pettijohn, *Jour. Geol.*, v. 58, pp. 169–171, 1950.

Turbidity currents and non-graded deposits, Ph. H. Kuenen and H. W. Menard, *Jnl. Sed. Petr.*, v. 22, pp. 83–96.

Turbidity currents and sediments in the North Atlantic, D. B. Ericson, M. Ewing and B. C. Heezen, *Amer. Assoc. Petr. Geol. Bull.*, v. 36, pp. 489–511, 1952.

Turbidity currents and submarine slumps and the 1929 Grand Banks earthquake, B. C. Heezen and M. Ewing, *Amer. Jour. Sci.*, v. 250, pp. 849–873, 1952.

Turbidity currents as a cause of graded bedding, Ph. H. Kuenen and C. I. Migliorini, *J. Geol.*, v. 58, pp. 91–127, 1950.

The unconsolidated sediments, J. I. Ewing and J. E. Nafe, *The Sea*, v. 3, 1963: Interscience, New York, pp. 73–84.

Chapter Seven
I Wasn't Lonely a Bit

Gravity anomalies and structure of the West Indies, Part I, M. Ewing and J. L. Worzel, *Geol. Soc. Amer. Bull.*, v. 65, pp. 165–174, 1954.

A letter to my children, W. M. Ewing, *The Reader's Digest*, Sept. 1954.

Neopolina (Vema) ewingi, a second living species of the Paleozoic class Monoplacophora, A. H. Clarke, Jr., and R. J. Menzies, *Science*, v. 129, pp. 1026–1027, 1959.

Oceanographic research programs of the Lamont Geological Observatory, M. Ewing and B. C. Heezen, *Geog. Rev.*, v. 46, pp. 508–535, 1956.

On the antiquity of the deep-sea bottom fauna, R. J. Menzies and J. Imbrie, *Oikos*, v. 9, pp. 192–210, 1958.

Polar wandering and climate, M. Ewing and W. L. Donn, *Polar Wandering and Continental Drift*, Soc. Econ. Paleo. and Mineral., pp. 94–99, 1963.

Precision measurement of ocean depth, B. Luskin, B. C. Heezen, M. Ewing and M. Landisman, *Deep-sea Res.*, v. 1, pp. 131–140, 1954.

Radiological studies in the investigation of ocean circulation, M. Ewing and R. D. Gerard, *Aspects of Deep Sea Research*, NAS-NRC Pub. 473, pp. 58–66, 1957.

Seismic profiler, J. I. Ewing and G. B. Tirey, *Jour. Geophys. Res.*, v. 66, pp. 2917–2927, 1961.

Strontium-90 in man, J. L. Kulp, W. R. Eckelmann, and A. R. Schulert, *Science*, v. 125, pp. 219–225, 1957.

A theory of ice ages, M. Ewing and W. L. Donn, *Science*, v. 123, pp. 1061–1066, 1956. *Ibid.*, v. 127, pp. 1159–1162, 1958. *Ibid.*, v. 152, pp. 1706–1712, 1966.

Whales entangled in deep-sea cables, B. C. Heezen, *Deep-sea Res.*, v. 4, pp. 105–115, 1957.

Chapter Eight
Rift and Ridge

Continuity of mid-oceanic ridge and rift valley in the southwestern Indian Ocean confirmed, M. Ewing and B. C. Heezen, *Science*, v. 131, pp. 1677–1679, 1960.

The deep-sea floor, B. C. Heezen, *Continental Drift*, 1962: Academic Press, New York.

Dynamic processes of abyssal sedimentation: Erosion, transportation, and redeposition on the deep-sea floor, B. C. Heezen, *Geophys. Jour. Roy. Astron. Soc.*, v. 2, pp. 142–163, 1959.

The floors of the oceans: I: The North Atlantic, B. C. Heezen, M. Tharp and M. Ewing, *Geol. Soc. Amer. Spec. Paper 65*.

The mantle rocks, J. I. Ewing, *The Sea*, v. 3, 1963: Interscience, New York, pp. 103–109.

The mid-oceanic ridge, B. C. Heezen and M. Ewing, *The Sea*, v. 3, 1963: Interscience, New York, pp. 388–410.

The mid-oceanic ridge and its extension through the Arctic Basin, B. C. Heezen and M. Ewing, *Geology of the Arctic*, 1961: Univ. Toronto Press, pp. 622–642.

Physiographic diagram of the South Atlantic Ocean, the Caribbean Sea, the Scotia Sea, and the eastern margin of the South Pacific Ocean, B. C. Heezen and M. Tharp, *Geol. Soc. Amer.*, 1961.

The rift in the ocean floor, B. C. Heezen, *Sci. Amer.*, v. 203, no. 4, pp. 98–110, 1960.

Some problems of Antarctic submarine geology, M. Ewing and B. C. Heezen, *Amer. Geophys. Union Geophys. Mon. 1*, pp. 75–81, 1956.

Topography of the deep-sea floor, B. C. Heezen and H. W. Menard, *The Sea*, v. 3, 1963: Interscience, New York, pp. 233–280.

World rift system, C. L. Drake, *Trans. Amer. Geophys. Union*, v. 45, pp. 435–440, 1964.

Chapter Nine
Diluvialists and Other Fauna

The Birth and Development of the Geological Sciences, F. D. Adams, 1954: Dover, New York.

Continental margins and geosynclines: The east coast of North America north of Cape Hatteras, C. L. Drake, M. Ewing and G. H. Sutton, *Physics and Chemistry of the Earth*, v. 3, 1959: Pergamon Press, London, pp. 110–198.

Darwin's Century, Loren Eiseley, 1958: Doubleday, New York.

The Founders of Geology, A. Geikie, Dover, New York.

Geophysical investigations in the emerged and submerged Atlantic coastal plain. Part V: Woods Hole, New York and Cape May sections, M. Ewing, J. L. Worzel, N. C. Steenland, and F. Press, *Geol. Soc. Amer. Bull.*, v. 61, pp. 877–892, 1950; Part VI: Long Island area, J. E. Oliver and C. L. Drake, *Geol. Soc. Amer. Bull.*, v. 62, pp. 1407–1417, 1951; Part VII: Continental shelf, continental slope and continental rise south of Nova Scotia, M. Ewing and C. B. Officer, *Geol. Soc. Amer. Bull.*, v. 65, pp. 653–670, 1954; Part VIII: Grand Banks and adjacent shelves, F. Press and W. Beckmann, *Geol. Soc. Amer. Bull.*, v. 65, pp. 299–314, 1954; Part IX: Gulf of Maine, C. L. Drake, J. L. Worzel, and W. C. Beckmann, *Geol. Soc. Amer. Bull.*, v. 65, pp. 957–970, 1954; Part X: Continental slope and continental rise south of Grand Banks, C. R. Bentley and J. L. Worzel, *Geol. Soc. Amer. Bull.*, v. 67, pp. 1–18, 1956.

Chapter Ten
To Gondwanaland

Ages of horizon A and the oldest Atlantic sediments, J. Ewing, J. L. Worzel, M. Ewing and C. Windisch, *Science*, v. 154, pp. 1125–1132, 1966.

Chain and Romanche fracture zones, B. C. Heezen, E. T. Bunce, and J. B. Hersey, *Deep-sea Res.*, v. 11, pp. 11–33, 1964.

Continent and ocean basin evolution by spreading of the sea floor, R. S. Dietz, *Nature*, v. 190, pp. 854–857, 1961.

Continental drift before 1900, N. A. Rupke, *Nature*, v. 227, pp. 349–350, 1970.

Debate About the Earth, H. Takeuchi, S. Uyeda, and H. Kanamori, 1967: W. H. Freeman, San Francisco.

Discussion, J. L. Worzel, Sympos. Continental Drift, *Phil. Trans. Roy. Soc.*, v. 258, pp. 137–139, 1965.

Distribution of oceanic sediments, M. Ewing and J. Ewing, *Studies in Oceanography*, Geophys. Inst., U. Tokyo, Japan, pp. 525–537, 1964.

Earth, An, S. K. Runcorn, *Nature*, v. 227, 1970, p. 525.

Earth, The, H. Jeffreys, 1970: Cambridge U. Press, London.

East Pacific Rise: The magnetic pattern and the fracture zones, M. Talwani, X. LePichon, and J. R. Heirtzler, *Science*, v. 150, pp. 1109–1115, 1965; World Rift Symposium, Ottawa, 1965.

Effect of earthquakes on deep-sea sediments, T. J. G. Francis, *Nature*, v. 233, pp. 98–102, 1971.

Field reversal or self-reversal, P. J. Smith, *Nature*, v. 229, pp. 378–380, 1971.

Fit of the continents around the Atlantic, E. C. Bullard, J. C. Everett, and A. G. Smith, *Phil. Trans. Roy. Soc. London*, A258, pp. 41–51.

Fit of the Southern Continents, A. G. Smith and A. Hallam, *Nature*, v. 223, pp. 139–144, 1970.

Geomagnetic reversals, A. Cox, *Science*, v. 163, pp. 237–244, 1969.

Geomagnetic reversals during the phanerozoic, M. W. McElhinny, *Science*, v. 172, pp. 157–159, 1971.

A geophysical study of the Red Sea, C. L. Drake and R. W. Girdler, *Geophys. Jour. Roy. Astron. Soc.*, v. 8, pp. 473–495, 1964.

History of Ocean Basins, H. Hess, *Petrologic Studies*, 1962: Geol. Soc. Amer., New York, pp. 599–620.

International Oceanographic Congress, M. Sears, ed., *Preprints*, 1959: AAAS, Washington; *Oceanography*, 1961: AAAS, Washington.

Magnetic anomalies over oceanic ridges, F. J. Vine and D. H. Matthews, *Nature*, v. 199, pp. 947–949, 1963.

Magnetic anomalies over the Reykjanes Ridge, J. R. Heirtzler, X. LePichon, and J. G. Baron, *Deep-sea Res.*, v. 13, pp. 427–443, 1966.

Magnetic evidence for horizontal displacements in the floor of the Pacific, V. Vacquier, *Continental Drift*, S. K. Runcorn, ed., 1962: Academic Press, New York.

The magnetism of the ocean floor, A. D. Raff, Sci. Amer., Oct. 1961, pp. 146–156.

A Mid-Labrador Sea ridge, C. L. Drake, N. J. Campbell, G. Sander, and J. E. Nafe, *Nature*, v. 200, pp. 1085–1086, 1963.

A new class of faults and their bearing on continental drift, J. T. Wilson, *Nature*, v. 207, p. 343, 1965.

Oceanic crustal studies, J. I. Ewing, *Trans. Amer. Geophys. Union*, v. 44, pp. 341–342, 1963.

The Origin of Continents and Oceans, A. Wegener, Dover, New York.

The origin of ocean ridges, Egon Orowan, *Sci. Amer.*, Nov. 1969, pp. 102–118.

Reversals of the earth's magnetic field, A. Cox, R. R. Doell, and G. B. Dalrymple, *Science*, v. 144, pp. 1537–1543, 1964.

Reversals of the earth's magnetic field, Allan Cox, G. B. Dalrymple, and R. R. Doell, *Sci. Amer.*, Feb. 1967, pp. 44–53.

A review of marine geophysics, M. Talwani, *Marine Geology*, v. 2, pp. 29–80, 1964.

Sediment distribution in the oceans: the Mid-Atlantic Ridge, M. Ewing, J. I. Ewing and M. Talwani, *Geol. Soc. Amer. Bull.*, v. 75, pp. 17–35, 1964.

Sediments of ocean basins, M. Ewing, *Man, Science, Learning and Education*, William Marsh Rice University, 1963, pp. 41–59.

Seismic-refraction measurements in the Atlantic Ocean basins, in the Mediterranean Sea, on the Mid-Atlantic Ridge, and in the Norwegian Sea, J. Ewing and M. Ewing, *Geol. Soc. Amer. Bull.*, v. 70, pp. 291–318, 1959.

Seismicity of the South Pacific Ocean, L. R. Sykes, *Jour. Geophys. Res.*, v. 68, pp. 5999–6006, 1963.

Submarine geophysics, M. Ewing, Amer. Geophys. Union, *Trans.*, v. 44, pp. 351–354, 1963.

The Vema fracture zone in the equatorial Atlantic, B. C. Heezen, R. D. Gerard, and M. Tharp, *Jour. Geophys. Res.*, v. 69, pp. 733–739, 1964.

Chapter Eleven
Eltanin-19

Crustal structure of the mid-ocean ridges. I: Seismic refraction measurements, X. LePichon, R. E. Houtz, C. L. Drake and J. E. Nave, *Jour. Geophys. Res.*, v. 70, pp. 319–339, 1965; II: Computed model from gravity and seismic refraction data, M. Talwani, X. LePichon, and M. Ewing, *ibid.*, pp. 341–352; III: Magnetic anomalies over the mid-Atlantic ridge, J. R. Heirtzler and X. LePichon, *ibid.*, pp. 4013–4033; IV: Sediment distribution. . . . , M. Ewing, X. LePichon and J. Ewing, *ibid.*, v. 71, pp. 1611–1636, 1966; V: Heat flow . . . and convection currents, M. G. Lanseth, X. LePichon, and M. Ewing, *ibid.*, v. 71, pp. 5321–5355, 1966.

Deep earthquake zones, anomalous structures in the upper mantle and the lithosphere, J. Oliver and B. Isacks, *Jour. Geophys. Res.*, v. 72, pp. 4259–4275, 1967.

Faunal extinctions and reversals of the earth's magnetic field, J. D. Hays, *Geol. Soc. Amer. Bull.*, v. 82, pp. 2433–2447, 1971.

Magnetic anomalies in the Indian Ocean and sea-floor spreading, X. LePichon and J. R. Heirtzler, *Jour. Geophys. Res.*, v. 73, pp. 2101–2117, 1967.

Magnetic anomalies in the Pacific and sea floor spreading, W. C. Pitman, E. M. Herron, and J. R. Heirtzler, *ibid.*, pp. 2069–2085.

Magnetic anomalies in the South Atlantic and ocean floor spreading, G. O. Dickson, W. C. Pitman, and J. R. Heirtzler, *ibid.*, pp. 2087–2100.

Magnetic anomalies over a young oceanic ridge off Vancouver Island, F. J. Vine and J. T. Wilson, *Science*, v. 150, pp. 485–489, 1965.

Magnetic anomalies over the Pacific Antarctic Ridge, W. C. Pitman and J. R. Heirtzler, *Science*, v. 154, pp. 1164–1171, 1966.

Magnetic boundaries in the North Atlantic Ocean, J. R. Heirtzler and D. E. Hayes, *Science*, v. 157, pp. 185–187, 1967.

The magnetic stratigraphy of a deep sea core from the North Pacific Ocean, G. O. Dickson and J. H. Foster, *Earth and Plan. Sci. Lett.*, v. 1, pp. 458–462, 1966.

Magnetism of the earth and climate changes, G. Wollin et al., *Earth and Plan. Sci. Lett.*, v. 12, pp. 175–183, 1971.

Mechanism of earthquakes and the nature of faulting on the mid-oceanic ridges, L. R. Sykes, *Jour. Geophys. Res.*, v. 72, pp. 2131–2153, 1967.

A paleomagnetic spinner magnetometer. . . . , J. H. Foster, *Earth and Plan. Sci. Lett.*, v. 1, pp. 463–466, 1966.

Paleomagnetic study of Antarctic deep-sea cores, N. D. Opdyke, B. Glass, J. D. Hays and J. Foster, *Science*, v. 154, pp. 349–357, 1966.

Paleomagnetism of deep-sea cores, N. D. Opdyke, *Rev. Geophys. and Space Phys.*, v. 10, pp. 213–249, 1972.

The Reykjanes Ridge crest: A detailed geophysical study, M. Talwani, C. C. Windisch, and M. G. Langseth, *Jour. Geophys. Res.*, v. 76, pp. 473–517, 1971.

Sea-floor spreading near the Galapagos, E. M. Herron and J. R. Heirtzler, *Science*, v. 158, pp. 775–780, 1967.

Sediment distribution on the mid-ocean ridges with respect to spreading of the sea floor, J. Ewing and M. Ewing, *Science*, v. 156, pp. 1590–1592, 1967.

Sediments and structure of the Japan trench, W. J. Ludwig et al., *Jour. Geophys. Res.*, v. 71, pp. 2121–2137, 1966.

Spreading of the ocean floor: New evidence, F. J. Vine, *Science*, v. 154, pp. 1405–1415, 1966.

Tektites and geomagnetic reversals, B. Glass and B. C. Heezen, *Nature*, v. 214, pp. 372–374, 1967.

Tektites and geomagnetic reversals, B. P. Glass and B. C. Heezen, *Sci. Amer.*, July 1967, pp. 32–38.

Tertiary sediment from the East Pacific Rise, L. H. Burckle, J. Ewing, T. Saito and R. Leyden, *Science*, v. 157, pp. 537–540, 1967.

Tertiary sediment from the Mid-Atlantic Ridge, T. Saito, M. Ewing, and L. H. Burckle, *Science*, v. 151, pp. 1075–1079, 1966.

Transform faults, oceanic ridges, and magnetic anomalies southwest of Vancouver Island, J. T. Wilson, *Science*, v. 150, pp. 482–485, 1965.

Chapter Twelve
Global Tectonics

Alaskan earthquake of 1964 and Chilean earthquake of 1960: Implications for arctectonics, G. Plafker, *Jour. Geophys. Res.*, v. 77, pp. 901–925, 1972.

Ancient continental mantle beneath oceanic ridges, E. Bonatti, *Jour. Geophys. Res.*, v. 76, pp. 3825–3831, 1971.

Arthur Holmes: Originator of the spreading ocean floor hypothesis, A. A. Meyerhoff, *Jour. Geophys. Res.*, v. 73, pp. 6563–6565, 1968. Reply, R. S. Dietz, p. 6567; Reply, H. H. Hess, p. 6569.

Asymmetric sea-floor spreading south of Australia, J. K. Weissel and D. E. Hayes, *Nature*, v. 231, pp. 518–522, 1971.

Composition and evolution of the mantle and core, D. L. Anderson, C. Sammis, T. Jordan, *Science*, v. 171, pp. 1103–1112, 1971.

Continental Drift, A. A. Meyerhoff and C. Teichert, *Jnl. Geology*, v. 79, 1971.

Continental drift and reserves of oil and natural gas, D. H. Tarling, *Nature*, v. 243, pp. 277–279, 1973.

Convection plumes in the lower mantle, W. J. Morgan, *Nature*, v. 230, pp. 42–43, 1971.

Cordilleran tectonic transitions and motion of the North American Plate, P. J. Coney, *Nature*, v. 233, pp. 462–465, 1971.

A crustal section across the eastern Alps based on gravity and seismic refraction data, S. Mueller and M. Talwani, *Pure and Applied Geophys.*, v. 85, pp. 226–239, 1971.

Deep-sea drilling for scientific purposes: a decade of dreams, T. H. van Andel, *Science*, v. 160, pp. 1419–1424, 1968.

Deep sea drilling in the South Atlantic, A. E. Maxwell et al., *Science*, v. 166, pp. 1047–1058, 1970.

Did an ice cap break Gondwanaland?, D. I. Gough, *Jour. Geophys. Res.*, v. 75, pp. 4475–4477, 1970.

Did the Atlantic close and then re-open?, J. T. Wilson, *Nature*, v. 211, pp. 676–681, 1960.

Disposal of waste material in tectonic sinks, R. C. Bostrom and M. A. Sherif, *Nature*, v. 226, pp. 154–156, 1970.

Distribution of stresses in the descending lithosphere from a global survey, B. Isacks and P. Molnar, *Rev. Geophys. and Space Phys.*, v. 9, pp. 103–174, 1971.

Elevation of ridges and the evolution of the central eastern Pacific, J. G. Sclater, R. N. Anderson and M. L. Bell, *Jour. Geophys. Res.*, v. 76, pp. 7888–7915, 1971.

Evidence from islands of the spreading of ocean floors, J. T. Wilson, *Nature*, v. 197, pp. 536–538, 1963.

Formation of the Indian Ocean, M. W. McElhinny, *Nature*, v. 228, pp. 977–979, 1970.

Franciscan melanges as a model for eugeosynclinal sedimentation and underthrusting tectonics, K. J. Hsu, *Jour. Geophys. Res.*, v. 76, pp. 1162–1169, 1971.

Fragmentation of the Alpine orogenic belt by microplate dispersal, W. Alvarez, T. Cocozza, F. C. Wezel, *Nature*, v. 248, pp. 309–314, 1974.

Geodynamics project, *EOS*, v. 52, pp. 396–405, 1971.

The geophysical year, F. J. Vine, *Nature*, v. 227, pp. 1013–1017, 1970.

Global tectonics, W. R. Dickinson, *Science*, v. 168, pp. 1250–1259, 1970.

Gondwanaland, paleomagnetism and continental drift, D. H. Tarling, *Nature*, v. 229, pp 17–21, 1971.

Gravitational mechanism for sea-floor spreading, H. N. Pollack, *Science*, v. 163, pp. 176–177, 1969.

Gravity anomalies and crustal shortening in the eastern Mediterranean, P. D. Rabinnowitz and W. B. F. Ryan, *Tectonophysics*, v. 10, pp. 585–608, 1971.

Gulf of California: a result of ocean-floor spreading and transform faulting, R. L. Larson, H. W. Menard, and S. M. Smith, *Science*, v. 161, pp. 781–783, 1968.

JOIDES, Ocean drilling on the continental margin, *Science*, v. 150, pp. 709–716, 1965.

Late Cenozoic evolution of the Great Basin, western United States, as an ensialic interarc basin, C. H. Scholz, M. Barazangi, and M. L. Sbar, *Geol. Soc. Amer. Bull.*, v. 82, pp. 2979–2990, 1971.

Lithospheric plate motion, sea level changes and climatic and ecological consequences, J. D. Hays and W. C. Pitman, *Nature*, v. 246, pp. 18–21, 1973.

Location of the proto-Atlantic suture in the British Isles, P. J. Gunn, *Nature*, v. 242, pp. 111–112, 1973.

Mantle earthquake mechanisms and the sinking of the lithosphere, B. Isacks and P. Molnar, *Nature*, v. 223, pp. 1121–1124, 1969.

Marginal offsets, fracture zones and the early opening of the North Atlantic, X. LePichon and P. J. Fox, *Jour. Geophys. Res.*, v. 76, pp. 6294–6308, 1971.

Marine magnetic anomalies, geomagnetic field reversals, and motions of the ocean floor and continents, J. R. Heirtzler, G. O. Dickson, E. M. Herron, W. C. Pitman, and X. LePichon, *Jour. Geophys. Res.*, v. 73, pp. 2119–2136, 1967.

Mechanical properties and processes in the mantle, L. R. Sykes, R. Kay, and O. L. Anderson, *EOS*, v. 51, pp. 874–879, 1970.

Metallic ore deposits and continental drift, D. H. Tarling, *Nature*, v. 243, pp. 193–195, 1973.

Metamorphism in the Mid-Atlantic Ridge, A. Miyashiro, F. Shido and M. Ewing, *Phil. Trans. Roy. Soc. Lond.*, v. A268, pp. 589–603, 1971.

A model for plate tectonic evolution of mantle layers, W. R. Dickinson and W. C. Luth, *Science*, v. 174, pp. 400–404, 1971.

Mountain belts and the new global tectonics, J. F. Dewey and J. M. Bird, *Jour. Geophys. Res.*, v. 75, pp. 2625–2647, 1970.

Oceanic sediment volumes and continental drift, J. Gilluly, *Science*, v. 166, pp. 992–993, 1969.

Origin and development of marginal basins in the western Pacific, D. E. Karig, *Jour. Geophys. Res.*, v. 76, pp. 2542–2561, 1971.

Origin and emplacement of the ophiolite suite: Appalachian ophiolites in Newfoundland, J. F. Dewey and J. M. Bird, *Jour. Geophys. Res.*, v. 76, pp. 3179–3206, 1971.

Petrologic models for the Mid-Atlantic Ridge, A. Miyashiro, F. Shido, and M. Ewing, *Deep-sea Res.*, v. 17, pp. 109–123, 1970.

Plate tectonic regulation of faunal diversity and sea level: A model, J. W. Valentine and E. M. Moores, *Nature*, v. 228, pp. 657–659, 1970.

Plate tectonics and the evolution of the Alpine system, J. F. Dewey, W. C. Pitman, W. B. F. Ryan, and J. Bonnin, *Geol. Soc. Amer. Bull.*, v. 84, pp. 3137–3180, 1973.

Plate tectonics in geologic history, W. R. Dickinson, *Science*, v. 174, pp. 107–113, 1971.

Plate tectonics of the Red Sea and East Africa, D. P. McKenzie, D. Davies, P. Molnar, *Nature*, v. 226, pp. 243–248, 1970.

Pre-drift continental nuclei, P. M. Hurley and J. R. Rand, *Science*, v. 164, pp. 1229–1242, 1969.

Reconstruction of Pangaea, R. S. Dietz and J. C. Holden, *Jour. Geophys. Res.*, v. 75, pp. 4939–4956, 1970.

A revolution in earth science, J. T. Wilson, *Geotimes*, v. 13, pp. 10–16, 1968. An open letter to J. Tuzo Wilson, V. V. Beloussov, *ibid.*, pp. 17–19; A reply to V. V. Beloussov, *ibid.*, pp. 20–22.

Ridges and basins of the Tonga-Kernadec island arc system, D. E. Karig, *Jour. Geophys. Res.*, v. 75, pp. 239–254, 1970.

Rises, trenches, great faults and crustal blocks, W. J. Morgan, *Jour. Geophys. Res.*, v. 73, pp. 1959–1982, 1968.

Rotation of the Corsica-Sardinia microplate, W. Alvarez, *Nature*, v. 235, pp. 103–105, 1972.

Saharan (paleozoic) glaciation dated in North America, K. Burke and J. B. Waterhouse, *Nature*, v. 241, pp. 267–268, 1973.

Sea-floor spreading and continental drift, X. LePichon, *Jour. Geophys. Res.*, v. 73, pp. 3661–3697, 1968.

Sea-floor spreading as thermal convection, W. M. Elsasser, *Jour. Geophys. Res.*, v. 76, pp. 1101–1112, 1971.

Sea-floor spreading in the Gulf of Alaska, W. C. Pitman and D. E. Hayes, *Jour. Geophys. Res.*, v. 73, pp. 6571–6580, 1968.

Sea-floor spreading in the North Atlantic, W. C. Pitman and M. Talwani, *Geol. Soc. Amer. Bull.*, v. 83, pp. 619–646, 1972.

Seismicity of the Indian Ocean and a possible nascent island arc between Ceylon and Australia, L. R. Sykes, *Jour. Geophys. Res.*, v. 75, pp. 5041–5055, 1970.

Seismology and the new global tectonics, B. Isacks and J. Oliver, *Jour. Geophys. Res.*, v. 73, pp. 5855–5899, 1967.

Seismology of the moon, and implications on internal structure, origin, and evolution, M. Ewing et al., *Highlights of Astronomy*, pp. 155–172, 1971.

The so-called folded mountains, W. M. Elsasser, *Jour. Geophys. Res.*, v. 75, pp. 1615–1618, 1970.

Speech, R. S. Dietz, *Transactions*, Amer. Geophys. Union, v. 52, p. 541, 1971.

Stratigraphy and structure of northeastern Newfoundland bearing on drift in North Atlantic, M. Kay, *Am. Assn. Petr. Geol. Bull.*, v. 51, pp. 579–600, 1967.

Test of continental drift by comparison of radiometric ages, P. M. Hurley et al., *Science*, v. 157, pp. 495–500, 1967.

Topological inconsistency of continental drift on present-sized earth, R. Meservey, *Science*, v. 166, pp. 609–611, 1969.

Two types of mountain building, J. T. Wilson and K. Burke, *Nature*, v. 239, pp. 448–449, 1972.

Ultramafics and orogeny, E. Moores, *Nature*, v. 228, pp. 837–842, 1970.

Was the Hercynian orogenic belt of Europe of the Andean type?, A. Nicholas, *Nature*, v. 236, pp. 221–224, 1772.

World Rift System, Symposium, Zurich (J. T. Wilson), p. 281, *Tectonophysics*, v. 8, 1969.

World seismicity maps, M. Barazangi and J. Dorman, *Bull. Seism. Soc. Amer.*, v. 59, pp. 369–380, 1969.

World-wide correlation of Mesozoic magnetic anomalies, and its implications, R. L. Larson and W. C. Pitman, *Geol. Soc. Amer. Bull.*, v. 83, pp. 3645–3661, 1972.

Picture Credits

Courtesy of Allan W. H. Bé: *57.*

Stanley A. Kling, courtesy of the Deep Sea Drilling Project: *58.*

From *The Depths of the Sea* by Charles Wyville Thomson: *55 (top).*

From "A Report on Deep Sea Deposits" by John Murray and A. F. Renard, as published in *Report on the Scientific Results of the Voyage of H.M.S. Challenger,* London: Her Majesty's Stationery Office, 1891: *55 (bottom).*

Courtesy of the British Tourist Authority: *60.*

Courtesy of the Woods Hole Oceanographic Institution: *67, 113 (bottom).*

Photograph by Don Fay, Woods Hole Oceanographic Institution, © National Geographic Society: *75.*

Courtesy of J. L. Worzel: *87.*

Photograph by William Wertenbaker: *89.*

Photograph by Ph. D. Kuenen, from *Principles of Physical Geology* by Arthur Holmes, 2nd edition, Ronald Press, 1965: *112 (top).*

From *Marine Geology of the Pacific* by H. W. Menard. Copyright © 1964 by McGraw-Hill, Inc. Used with permission of the McGraw-Hill Book Company: *113 (top).*

From "A Geophysical Study of the Red Sea" by Charles L. Drake and R. W. Girdler, as published by the Royal Astronomical Society in the *Geophysical Journal,* vol. 8, no. 5, 1964: *152 (top).*

Photograph by Haroun Tazieff: *152 (bottom).*

Photograph by Enrico Bonatti: *150.*

Courtesy of the Princeton Museum of Natural History: *155.*

From *Geology Illustrated* by John S. Shelton. W. H. Freeman and Company. Copyright © 1966: *163.*

Courtesy of the Iceland Tourist Agency: *164.*

Courtesy of the National Air Photo Library of Canada: *172.*

Courtesy of British Antarctic Survey. Copyright by the Trans-Antarctica Association: *173.*

Adapted from "Fit of the Continents Around the Atlantic" by E. C. Bullard, J. C. Everett and A. G. Smith, from *Philosophical Transactions,* A. 258, as published by the Royal Society, London, 1965; and from "Fit of the Southern Continents" by A. G. Smith and A. Hallam, as published in *Nature,* vol. 225: 1970: *179.*

Index

abyssal gaps, 142
abyssal hills, 137, 140–141, 142, 143–144
abyssal plains, 78–80, 82, 84, 85; seismic profiles, 80; and Mid-Atlantic Ridge, 86, 109–110, 138–139; and turbidity currents, 119; and precision depth recorder, 127; maps of, in North Atlantic, 137; and continental rise, 139; and Puerto Rico Trench, 140; described, 140–144; and seismic reflection profiler, 183
acoustics, marine, 16, 40–41; during World War II, 41–43; Ewing's research at Woods Hole (World War II), 44, 98; and Hudson Laboratories, 91
Aden, 128
Africa, 142; and continental drift concept, 171, 174; and Mid-Ocean Ridge, 180, 181; and plate tectonics,
216, 224; pre-Cambrian sutures, 234
See also East Africa; West Africa
African Rift Valleys, 147, 177; and beginnings of continental breakup, 236
Agassiz, Alexander, 63–64
Agassiz, Louis, 52
Alaska, 153, 218, 220, 227
Albatross, Swedish expedition, 78
Albatross Plateau, west of Central America, 147
Albertus Magnus, 155–156
Aleutian Trench, 220
Alps, and plate tectonics, 210, 220, 224
aluminum, 141
Alvin, research submarine, 30
Amazon River, submarine canyon, 107
American Geophysical Union, 22, 23, 203, 216, 218; and "Mid-Atlantic Seismic Belt" (lecture), 147
Anchorage, Alaska, 220
Andes, and plate tectonics, 210